AGAINST
CIVILIZATION

AGAINST
CIVILIZATION

READINGS AND REFLECTIONS
ENLARGED EDITION

EDITED BY JOHN ZERZAN

ILLUSTRATIONS BY R.L. TUBBESING

FERAL HOUSE

Major thanks to Alice Parman for assistance and advice concerning every aspect of this book and to Kevin Tucker whose invaluable help made the second edition a reality.

A Feral House Book
ISBN: 0-922915-98-9
ISBN: 978-0-9229-1598-9

Feral House
1240 W. Sims Way #124
Port Townsend, WA 98368

10 9 8 7 6 5

www.feralhouse.com
info@feralhouse.com

Illustrations © R.L. Tubbesing

Designed by Hedi El Kholti

Contemporary records indicate that, more than once, both rich and poor wished that the barbarians would deliver them from the [Roman] Empire. While some of the civilian population resisted the barbarians (with varying degrees of earnestness), and many more were simply inert in the presence of the invaders, some actively fought for the barbarians. In 378, for example, Balkan miners went over en masse to the Visigoths. In Gaul the invaders were sometimes welcomed as liberators from the Imperial burden, and were even invited to occupy territory.

—Joseph Tainter

To combat cultural genocide one needs a critique of civilization itself.

—Gary Snyder

VISUALIZE INDUSTRIAL COLLAPSE

—Earth First!

CONTENTS

PAGE

Foreword: Chellis Glendinning xi

Introduction: John Zerzan 1

Preface: Kevin Tucker, *Unintended Consequences* 4

SECTION I. OUTSIDE CIVILIZATION 8

   Roy Walker, *The Golden Feast* 11

   Hoxie Neale Fairchild, *The Noble Savage: A Study in Romantic Naturalism* . . . 15

   Jean-Jacques Rousseau, *Discourse on the Origins of Inequality* 19

   Henry David Thoreau, "Excursions" 25

   Fredy Perlman, *Against His-story, Against Leviathan!* 27

   Arnold DeVries, *Primitive Man and His Food* 31

   Marshall Sahlins, "The Original Affluent Society" 35

   Lynn Clive, "Birds Combat Civilization" 39

   John Landau, "Wildflowers: A Bouquet of Theses" 41

   Theodor Adorno, *Minima Moralia: Reflections from Damaged Life* 44

   Marvin Harris, *Our Kind* 46

   Ramona Wilson, *Spokane Museum* 49

SECTION II. THE COMING OF CIVILIZATION 50

   George P. Marsh, *The Earth as Modified by Human Action* 53

   Frederick Turner, *Beyond Geography: The Western Spirit Against the Wilderness* . . 58

   James Axtell, *The Invasion Within: The Contest of Cultures in Colonial North America* . 65

   John Zerzan, *Elements of Refusal* 68

   Paul Shepard, *Nature and Madness* 74

   Mark Nathan Cohen, *Health and the Rise of Civilization* 81

   Robin Fox, *The Search for Society* 90

   Chellis Glendinning, *My Name is Chellis and I'm in Recovery from Western Civilization* 91

   Pierre Clastres, *Society Against the State* 95

   Madhusree Mukerjee, *The Land of the Naked People* 97

   Robert Wolff, *Reading and Writing* 99

Section III. The Nature of Civilization 102

    Friedrich Schiller, *On the Aesthetic Education of Man* 105

    Charles Fourier, *Theory of Four Movements and General Destinies* . . . 107

    Sigmund Freud, *Civilization and Its Discontents* 108

    John Landau, "Civilization and the Primitive" 110

    Max Horkheimer, *Eclipse of Reason* 112

    Max Horkheimer, *Dawn and Decline* 115

    Richard Heinberg, "Was Civilization a Mistake?" 116

    Barbara Mor, *Here: a small history of a mining town in the american southwest:* . . 124
        warren/bisbee az . . . .

    Ivan Illich, *Toward a History of Needs* 129

    Zygmunt Bauman, *Modernity and the Holocaust* 131

    T. Fulano, "Civilization Is Like a Jetliner" 136

    Unabomber (AKA "FC"), "Industrial Society and Its Future" . . . 139

    Tamarack Song, *The Old Way and Civilization* 141

    Ursula K. LeGuin, *Women/Wilderness* 147

Section IV. The Pathology of Civilization 150

    Max Nordau, *Conventional Lies, or Our Civilization* 153

    William H. Koetke, *The Final Empire: The Collapse of Civilization* . . . 155
        and The Seed of the Future . . . .

    Joseph A. Tainter, *The Collapse of Complex Societies* 162

    Theodore Roszak, *Where the Wasteland Ends: Politics and Transcendence* . . 167
        in Postindustrial Society . . . .

    Andrew Bard Schmookler, *The Parable of the Tribes: The Problem* . . 172
        of Power in Social Evolution . . .

    Peter Sloterdijk, *Critique of Cynical Reason* 176

    Fredric Jameson, *The Seeds of Time* 179

    labor of ludd, "The Medium Is the Medium" 182

    Des Réfractaires, "How Nice to Be Civilized!" 184

    David Watson, "Civilization in Bulk" 187

    Richard Heinberg, *Memories and Visions of Paradise* 196

    Chrystos, "They're Always Telling Me I'm Too Angry" 198

    Oswald Spengler, *Man and Technics: A Contribution to a Philosophy of Life* . . 204

    John Mohawk, *In Search of Noble Ancestors* 206

SECTION V. THE RESISTANCE TO CIVILIZATION 212

Rudolf Bahro, *Avoiding Social and Ecological Disaster: The Politics of World Transformation* 215

John Zerzan, *Future Primitive* 220

William Morris, *News from Nowhere* 222

Feral Faun, "Feral Revolution" 227

Anonymous, "Don't Eat Your Revolution! Make It!" 231

Glenn Parton, "The Machine in Our Heads" 232

Alon K. Raab, "Revolt of the Bats" 243

Kirkpatrick Sale, *Rebels Against the Future: Lessons from the Luddites* . . . 247

Derrick Jensen, "Actions Speak Louder Than Words" 252

Anti-Authoritarians Anonymous, "We Have To Dismantle All This" . . . 256

John (Fire) Lame Deer and Richard Erdoes, *Talking to the Owls and Butterflies* . 258

Group of Anarcho-Futurists, *Anarcho-Futurist Manifesto* 263

Susan Griffin, *Woman and Nature: The Roaring Inside Her* 265

Communique #23 from *Disorderly Conduct* #6, "Why Civilization?" . . . 266

Sources 273

FOREWORD

The thing I admire about Chicano village life in New Mexico is that beneath the sleek overlay of trucks and telephones lies the still-vital infrastructure of an ancient and, until recently, undisturbed way of life. Men hunt elk and turkey. Women know plants. Curandera-healers with their potent prophetic powers live among us. Everyone knows how to build a mud house, dig the irrigation ditch, grow corn, ride a horse, and navigate through the forest on a moon-lit night. And despite the raging poverty that universally flattens land-based communities when they are conquered, colonized, and consumed, there is more happiness here than in any place I have known. It's a simple happiness, nothing fancy, a here-and-now contentment: a story told at the gas pump, an invitation to go fishing, a twist of language that illuminates the irony of history. Living here, I have learned not to contemplate a trip to the dump or the village store without carrying with me an extra twenty minutes, or an hour, to give and receive.

Such experiences reinforce what, after decades of research and dozens of social-change movements, I have long suspected. That it is not just contemporary industrial society that is dysfunctional; it is civilization itself. That we humans are born to be creatures of the land and the sea and the stars; that we are relations to the animals, cohorts to the plants. And that our well-being, and the well-being of the very planet, depend on our pursuance of our given place within the natural world.

It is against these musings that I celebrate the coming of John Zerzan's accomplishment of an anthology harboring the best of civilized people's critiques of civilization. Herein the reader will discover the questions that need to be asked and the insights that beg to be nurtured if humankind and the natural world as we know it are to thrive into the future.

This book is that important.

—Chellis Glendinning
Chimayó, New Mexico
26 July 1998

INTRODUCTION

JOHN ZERZAN

Since the first edition of *Against Civilization* [1999] the general perspective referred to by its title has begun to make sense to a growing number of people. An overall crisis—personal, social, environmental—is rapidly deepening, making such an indictment feasible, if not unavoidable.

This collection is, among other things, a reminder that critiques of civilization itself are anything but new. And the past five years have provided an opportunity to add voices to the chorus of doubters, those with enough vision to think outside civilizational confines.

The 15 additional selections include correctives to a serious deficiency of the original book: the paucity of women and indigenous contributors [not exclusive categories, of course].

The current edition makes some advance in these vital areas, I believe.

Discontent with civilization has been with us all along, but is coming on now with a new freshness and insistence, as if it were a new thing. To assail civilization itself would be scandalous, but for the conclusion, occurring to more and more people, that it may be civilization that is the fundamental scandal.

I won't dwell here on the fact of the accelerating destruction of the biosphere. And perhaps equally obvious is the mutilation of "human nature," along with outer nature. Freud decided that the fullness of civilization would bring, concomitantly, the zenith of universal neurosis. In this he was evidently a bit sanguine, too mild in his prognosis.

It is impossible to scan a newspaper and miss the malignancy of daily life. See the multiple homicides, the 600-percent increase in teen suicide over the past 30 years; count the ways to be heavily drugged against reality; ponder what is behind the movement away from literacy. One could go on almost endlessly charting the boredom, depression, immiseration.

The concept of progress has been in trouble for a few decades, but the general crisis is deepening now at a quickening pace. From this palpable extremity it is clear that something is profoundly wrong. How far back did this virus originate? How much must change for us to turn away from the cultural death march we are on?

At the same time, there are some who cling to the ideal of civilization, as to a promise yet to be fulfilled. Norbert Elias, for example, declared that "civilization is never finished and always endangered." More persuasive is the sobering view of what civilization has already wrought, as in today's deadening and deadly convergence of technological processes and mass society. Richard Rubenstein found that the Holocaust "bears witness to the advance of civilization," a chilling point further developed by Zygmunt Bauman in his *Modernity and the Holocaust*. Bauman argued that history's most gruesome moment so far was made possible by the inner logic of civilization, which is, at bottom, division of labor. This division of labor, or specialization, works to dissolve moral accountability as it contributes to technical achievement in this case—to the efficient, industrialized murder of millions.

But isn't this too grim a picture to account for all of it? What of other aspects, like art, music, literature—are they not also the fruit of civilization? To return to Bauman and his point about Nazi genocide, Germany was after all the land of Goethe and Beethoven, arguably the most cultural or spiritual European country. Of course we try to draw strength from beautiful achievements, which often offer cultural criticism as well as aesthetic uplift. Does the presence of these pleasures and consolations make an indictment of the whole less unavoidable?

Speaking of unfulfilled ideals, however, it is valid to point out that civilization is indeed "never finished and always endangered." And that is because civilization has always been imposed, and necessitates continual conquest and repression. Marx and Freud, among others, agreed on the incompatibility of humans and nature, which is to say, the necessity of triumph over nature, or work.

Obviously related is Kenneth Boulding's judgment that the achievements of civilization "have been paid for at a very high cost in human degradation, suffering, inequality, and dominance."

There hasn't been unanimity as to civilization's most salient characteristic. For Morgan it was writing; for Engels, state power; for Childe, the rise of cities. Renfrew nominated insulation from nature as most fundamental. But domestication stands behind all these manifestations, and not just the taming of animals and plants, but also the taming of human instincts and freedoms. Mastery, in various forms, has defined civilization and gauged human achievement. To name, to number, to time, to represent—symbolic culture is that array of masteries upon which all subsequent hierarchies and confinements rest.

Civilization is also separation from an original wholeness and grace. The poor thing we call our "human nature" was not our first nature; it is a pathological condition. All the consolations and compensations and prosthetics of an ever more technicized and barren world do not make up for the emptiness. As Hilzheimer and others came to view domestications of animals as juvenilizations, so also are we made increasingly dependent and infantilized by the progress of civilization.

Little wonder that myths, legends, and folklore about gardens of Eden, Golden Ages, Elysian fields, lands of Cockaigne, and other primitivist paradises are a worldwide phenomenon. This universal longing for an aboriginal, unalienated state has also had its dark flip side, a remarkable continuity of apocalyptic beliefs and prophets of doom—two sides of the same coin of a deep unhappiness with civilization.

Centuries of the persistence of utopias in the literature and politics of the West have more recently been replaced by a strong dystopian current, as hope seems to be giving way to nightmare apprehensions. This shift began in earnest in the nineteenth century, when virtually every major figure—e.g., Goethe, Hegel, Kierkegaard, Melville, Thoreau, Nietzsche, Flaubert, Dostoevski—expressed doubt about the vitality and future of culture. At the time that technology was becoming a worldwide unifying force, social scientists such as Durkheim and Masaryk noted that melancholy and suicide increased precisely with the forward movement of civilization.

In terms of the current intellectual domestication, postmodernism, despite a certain rhetoric of rebellion, is merely the latest extension of the modern civilizing process. For its moral cowardice as well as its zero degree of content, a horrific present is thus captured all too well. Meanwhile, *Forbes* magazine's 75th-anniversary cover story explored "Why We Feel So Bad When We Have It So Good," and the simple graffito "I can't breathe!" captures our contemporary reality with precision.

From every camp, voices counsel that there can be no turning back from the path of progress, the unfolding of still more high-tech consumerist desolation. How hollow they sound, as we consider what has been lost and what may yet, one desperately hopes, be recovered.

Preface

Unintended Consequences

Kevin Tucker

For millions of years, humans have lived as anarchists. That is as autonomous individuals without the existence of coercive power, work, and institutions: without mediation. The "state of nature" may more appropriately be called the natural anti-state. It was never paradise (the walled gardens) or utopia (the perfect place of the imagination), it just was. But it's not simply a historical thing either. The Linear thought of Reason would have us believe so, being led by the prophets of Production (Moses, Smith, Marx, etc.). Anarchy is in our bones. It's the way we act; it's the way that millions of years of evolution have shaped us. As Paul Shepard puts it, we are beings of the paleolith: gatherer-hunters, primitives, beings of this earth.

But something happened. This is no great mystery and no matter how devoted we are to the gods of Progress and Production, we all know things aren't going too great. We've been led astray. To try and confront what that means, we must first understand what we are. The life of nomadic gatherer-hunters is intrinsically different from the spiritually dead world of modernity: the current face of the global, technological civilization. Gatherer-hunters themselves are no different though. There are no born primitives or born civilized, but people born of different time and place and a large portion of us who had the misfortune of being born in the latter category.

Nomadic gatherer hunter societies typify egalitarianism. They are, as they must be by their nature, flexible and organic. Being nomadic means being adaptive: that is the key to anarchy. When there are droughts, societies can move to more hospitable regions. Boundaries, where they exist at all, are defined by the center rather than arbitrary lines or markers. Who is at one site at any particular time is fluid and there are no strangers. Egos are intentionally deflated so that no skill becomes more valued than others. Population is kept in check by the nature of mobility and what Richard B. Lee calls the "contraceptive on the hip."

But most importantly, everyone is capable of fully sustaining themselves. So when people do group together, they are doing so on their own terms. If people get angry or frustrated with others, they are free to leave and the impact of being shunned is harshly felt. There are no real specialists and no possessions that cannot be made or exchanged easily. *There is no mediation between life and means of living.*

The nomadic gatherer-hunters live in an entirely sacred world. Their spirituality reaches as far as all of their relations. They know the animals and plants that surround them and not only ones of immediate importance. They speak with what we would call "inanimate objects", but they can speak the same language. They know how to see beyond themselves and are not limited to the human languages that we hold so dearly. Their existence is grounded in place, they wander freely, but they are always home, welcome and fearless.

It's easy to criticize any theory that looks at "original sins" or points fingers towards any particular event. In many ways I agree, but I think the picture is really more complex. At no point was there a conscious decision to be civilized or a point when people stopped listening to the earth. Instead there are certain things that have happened and have had serious implications for the way we deal with each other and the earth around us.

I don't think that the first people to domesticate plants and animals knew what they did would turn the world they loved into something to eventually fear. Or that growing the fear of wildness would eventually mean destroying everything outside the barriers of the gardens to ensure that they did not creep in. It's really doubtful that the first people to settle in one area thought they were taking steps toward a life of warfare. Or that having more children would mean a constant and increasing state of growth. It's doubtful that the first people to become largely dependent upon stored foods would realize that this would mean the creation of coercive power and break the egalitarianism that a group of autonomous people had.

Of course, none of us will ever know for sure what was being thought or why these things were being done. There's no shortage of theories about the origins of domestication, sedentism or surplus-orientation, but for all practical purposes those theories are really irrelevant. Why steps were taken in the first place does not change the fact that those steps have carried a number of implications. When each of those steps was taken, something significant did happen and a trail of unintended consequences connects those events with where we are now.

But this is not a sign that governments or power is merely some benign force. Politicians and profiteers know that they are destroying the planet and poisoning all life, they just see money as more important. Their choices are hardly "unintended" compared to the person who unthinkingly plugs into an electrical outlet or pours gasoline into a car. Powermongers will act in their own interest, but their power relies on our complacency with the terms they have poured us in.

This doesn't mean that every person involved is necessarily aware or that they should be damned; that doesn't get us very far. But what is obvious is that our situation is getting increasingly worse. With the growing dependency on fossil fuels, we are stealing from the future in a way never known before. We are standing in a rather familiar position: like the Cahokia, Chacoan, Maya, Aztec, Mesopotamian, and Roman civilizations before us, we aren't seeing the symptoms of collapse that define our times. We aren't thinking about anything but what is good for us here and now. We aren't thinking outside of our conditioning. *We aren't thinking outside of civilization.*

But we don't even know it. We aren't even given the ability to read the times, because it is contrary to the Rational path of Reason laid out before us.

But things have changed and are changing. Whether we recognize it or not: *something will happen.* We have the ability to look back and try and awaken the part of ourselves that has been buried by domestication: the civilizing process. We can see that there is something about the nomadic gatherer-hunter existence that just worked. We can see that this was broken down by sedentism, domestication, surplus, and those breakdowns would solidify further with horticulture, the creation of states, agriculture and even more so with industrialism and technological modernity.

Something about these steps took away our autonomy. They made us dependent. Supposedly we were freed from the barbarism of self-determination toward the new Freedom of work and a world of stuff. We sold egalitarianism for plastic.

Our current situation is a grim one but we are not hopeless. We have before us the legacy of unintended consequences that slowly took us from egalitarianism to totalitarianism. The question we have to ask is what have we lost. What part of our being has been sold off in the process? We can look beyond the myths of Reason, of divine, Linear time, and Progress, and awaken ourselves in the process.

Civilization is a huge target. Overcoming domestication is a massive undertaking, but our souls and lives are at stake. But the future and the past are closer than we think. The blood and spirit of anarchy flows through our veins. We don't need to look 'before civilization'; we just need to listen to ourselves and the world around us. We have the benefit of seeing what steps have taken us down the wrong path, and with that we can start taking steps toward anarchy.

And in this process, the process of *becoming human*, the abstractions between our fate and the fate of the world will wither. There will be no question of when it is the right time to strike against the concrete manifestations of civilization or to know where to strike.

When we learn to open ourselves to wildness and chaos, the organic anarchy of our beings will flow. Attacking civilization is no easy feat, but when we listen, when we embrace our anti-state of nature, we will know exactly what to do.

—Kevin Tucker
10 May 2004

SECTION I

—

OUTSIDE

CIVILIZATION

Vyaghra

Neandertals did not paint their caves with the images of animals. But perhaps they had no need to distill life into representations, because its essences were already revealed to their senses. The sight of a running herd was enough to inspire a surging sense of beauty. They had no drums or bone flutes, but they could listen to the booming rhythms of the wind, the earth, and each other's heartbeats, and be transported.

—James Shreeve (1995)

This collection opens with some reflections about what it was like for our species prior to civilization.

In a literary vein, the pages from Roy Walker's classic treasury of poetry, *Golden Feast* (1952), remind us that from Ovid to the American Big Rock Candy Mountain folk legend, the memory or vision of an uncorrupted original wholeness persists. In fact, utopian anti-civilization longings reach back at least as far as the earliest Greek writings. From Hesiod's *Works and Days*, dating from the early seventh century B.C., came the canonical description of the Golden Age, the bitterly lamented vanished epoch of Kronos' reign, when humans "lived as if they were gods, their hearts free from sorrow, and without hard work or pain," when "the fruitful earth yielded its abundant harvest to them of its own accord, and they lived in ease and peace upon the lands with many good things."

Obviously this refers to the vast Paleolithic era, comprising more than 99 percent of our time span as a species. Current anthropology tells us that the pre-agricultural foraging life did not know organized violence, sexual oppression, work as an onerous or separate activity, private property, or symbolic culture. Reworked by Virgil and Ovid as the lost age of Saturn (the Roman Kronos), Hesiod's Golden Age reappeared as Arcadia, and the idyll has persisted in cultures everywhere. Richard Heinberg's *Memories and Visions of Paradise* (1995) is, by the way, an unexcelled recent exploration of this theme.

Fairchild's eminent study *Noble Savage* (1928) introduces the innocence of native New World peoples, soon to be lost to disease and warfare, upon the arrival of early conquerors. Rousseau, the origin of Fairchild's title, describes the felicity and freedom that once obtained.

The excerpt from Thoreau is a brief but lively one: "the most alive is the wildest," is his heartfelt conclusion. Perlman's intensity, in his superb *Against His-story, Against Leviathan* (1983), leaves little doubt as to the nature-based authenticity of those not subdued by civilization, as seen in their sense of play and autonomy, for example.

DeVries summarizes features of nondomesticated robustness and vitality in sharp contrast to later degeneracy in health. Sahlins' offering is an early statement of the central point of his *Stone Age Economics* (1972), namely, that paleolithic peoples are truly affluent, with no artificially produced or unmet needs.

Lynn Clive objects to the sacrifice of birds to skyscrapers and jetliners, while Landau offers a personal response to all we have lost. In a marvelous meditation, Adorno describes the utopian component of children's make-believe play. He recalls the pretamed stage of humanity in which productivity as a value is clearly refused, and exchange disregarded, as such nonutilitarian activity "rehearses the right life."

Ramona Wilson's moving poem and Marvin Harris' questioning of the inevitability of hierarchy augment the section.

THE GOLDEN FEAST (1952)

ROY WALKER

The fullest Roman expression of the Golden Age theme is in Ovid, a poet who completed his education at Athens. The last and greatest book of the *Metamorphoses* is devoted to the Pythagorean philosophy, and bears that title. In Dryden's translation this final book is the starting point of our endeavour to trace this tradition through the eighteenth century, and although the poem is a Roman achievement we may defer consideration of it. Ovid's first book deals with the grandest metamorphosis of all, the transformation from the Chaos that preceded Nature's birth to the comparative order of Caesar's time. In that great change an empire greater than Caesar's is won and lost, a Golden Age of peace and plenty, lost to be found again by those who carry a vision of it through darkness and observe its precepts of peace and harmlessness to all that lives. This is the golden legend that has haunted the imagination of Europe's prophets, regardless of their own temperaments, habits or cultural environment. In essentials it is also the story of Genesis and its history is inevitably joined with that of the first book of the Bible.

> Then sprang up first the golden age, which of itself maintained
> The truth and right of everything, unforced and unconstrained.
> There was no fear of punishment, there was no threatening law
> In brazen tables naile'd up, to keep the folk in awe.
> There was no man would crouch or creep to judge with cap in hand;
> They live'd safe without a judge in every realm and land.
> The lofty pine-tree was not hewn from mountains where it stood,
> In seeking strange and foreign lands to rove upon the flood.
> Men knew none other countries yet than where themselves did keep:
> There was no town enclose'd yet with walls and ditches deep.
> No horn nor trumpet was in use, no sword nor helmet worn.
> The world was such that soldiers' help might easily be forborne.
> The fertile earth as yet was free, untouched of spade or plough,
> And yet it yielded of itself of every thing enow;
> And men themselves contented well with plain and simple food

That on the earth by Nature's gift without their travail stood,
Did live by raspis, hips and haws, by cornels, plums and cherries,
By sloes and apples, nuts and pears, and loathsome bramble berries,
And by the acorns dropped on ground from Jove's broad tree in field.
The springtime lasted all the year, and Zephyr with his mild
And gentle blast did cherish things that grew of own accord.
The ground untilled all kind of fruits did plenteously afford.
No muck nor tillage was bestowed on lean and barren land
To make the corn of better head and ranker for to stand.

Then streams ran milk, then streams ran wine, and yellow honey flowed
From each green tree whereon the rays of fiery Phoebus glowed.
But when that unto Limbo once Saturnus being thrust,
The rule and charge of all the world was under Jove unjust,
And that the silver age came in, more somewhat base than gold,
More precious yet than freckled brass, immediately the old
And ancient springtime Jove abridged and made thereof anon
Four seasons: winter, summer, spring, and harvest off and on.
Then first of all began the air with fervent heat to swelt;
Then icicles hung roping down; then, for the cold was felt,
Men 'gan to shroud themselves in house; their houses were the thicks,
And bushy queaches, hollow caves, or hurdles made of sticks.
Then first of all were furrows drawn, and corn was cast in ground;
The simple ox with sorry sighs to heavy yoke was bound.

Next after this succeeded straight the third and brazen age:
More hard of nature, somewhat bent to cruel wars and rage,
But yet not wholly past all grace.
                              Of iron is the last
In no part good and tractable as former ages past;
For when that of this wicked age once opened was the vein
Therein all mischief rushéd forth, the faith and truth were fain
And honest shame to hide their heads; for whom stepped stoutly in,
Craft, treason, violence, envy, pride, and wicked lust to win.
The shipman hoists his sails to wind, whose names he did not know;
And ships that erst in tops of hills and mountains high did grow,
Did leap and dance on uncouth waves; and men began to bound
With dowls and ditches drawn in length the free and fertile ground,
Which was as common as the air and light of sun before.
Not only corn and other fruits, for sustenance and for store,
Were now exacted of the earth, but eft they 'gan to dig

And in the bowels of the earth insatiably to rig
For riches couched, and hidden deep in places near to hell,
The spurs and stirrers unto vice, and foes to doing well.
Then hurtful iron came abroad, then came forth yellow gold
More hurtful than the iron far, then came forth battle bold
That fights with both, and shakes his sword in cruel bloody hand.
Men live by ravin and by stealth; the wandering guest doth stand
In danger of his host; the host in danger of his guest;

And fathers of their sons-in-law; yea, seldom time doth rest
Between born brothers such accord and love as ought to be;
The goodman seeks the goodwife's death, and his again seeks she;
With grisly poison stepdames fell their husbands' sons assail;
The son inquires aforehand when his father's life shall fail;
All godliness lies under foot. And Lady Astrey, last
Of heavenly virtues, from this earth in slaughter drownéd passed.

Ovid's lines recreate the vision of the Ages of Gold, Silver, Brass and Iron, set down some seven hundred years before by Hesiod in *Works and Days*. Captured Greece, as the candid Horace says, had captured her rough conqueror.

In Hesiod's Golden Age, the first beatitude is the tranquil mind which, rather than a high material standard of living, is the highest good. Freedom from toil, next celebrated, expressed man's harmonious place in the natural order, in contrast to our civilization's war on soil, animal and tree. Long life, free from violence and disease, is as natural to the Golden Age as the abundance of fruits on which mankind is nourished there. All things are shared. All men are free.

We have vestigial modern doctrines for all these qualities: pacifism, vegetarianism, communitarianism, anarchism, soil conservation, organic farming, "no digging," afforestation, nature cure, the decentralised village economy. At the golden touch of Hesiod's or Ovid's lines the clumsy polysyllables crack their seed cases and flower into the variegated life and colour of single vision. The vague association that many of these ideas have retained in their attenuated modern forms is not accidental....

Finally, we may notice what seems to be the American's own version of Cockaigne, *The Big Rock Candy Mountains*:

One evening as the sun went down
And the jungle fire was burning

Down the track came a hobo hiking,
And he said "Boys I'm not turning.
I'm headed for a land that's far away,
Beside the crystal fountains,
So come with me, we'll all go and see
The big Rock Candy Mountains.

In the big Rock Candy Mountains,
There's a land that's fair and bright,
Where the hand-outs grow on bushes,
And you sleep out every night.
Where the box cars are all empty,
Where the sun shines every day,
On the birds and the bees,
And the cigarette trees,
And the lemonade springs
Where the blue-bird sings,
In the big Rock Candy Mountains.

In the big Rock Candy Mountains,
All the cops have wooden legs,
The bull-dogs all have rubber teeth
And the hens lay soft-boiled eggs.
The farmer's trees are full of fruit
And the barns are full of hay.
Oh I'm bound to go
Where there ain't no snow,
Where they hung the Turk
That invented work,
In the big Rock Candy Mountains.

In the big Rock Candy Mountains
You never change your socks.
And the little streams of alcohol
Come trickling down the rocks.
Where the brakemen have to tip their hats,
And the rail-road bulls are blind.
There's the lake of stew,
And of whisky too.
You can paddle all around 'em
In a big canoe
In the big Rock Candy Mountains.

THE NOBLE SAVAGE: A STUDY IN ROMANTIC NATURALISM (1928)

HOXIE NEALE FAIRCHILD

The narratives of Columbus illustrate the first step in the forma-tion of the Noble Savage idea. The Caribs are represented as a virtuous and mild people, beautiful, and with a certain natural intel-ligence, living together in nakedness and innocence, sharing their property in common. But though Columbus is enthusiastic about the Indians, he does not compare them with the Europeans. For such a comparison a stimulus was soon provided by the brutality of the Spaniards. Humanitarianism is the motive back of the *Breuisima Relación de la Destruyción de las Indias* of Las Casas.

By 1539, when Las Casas' book appeared, Spanish goldlust had made oppressed slaves of the free and amiable beings described by Columbus. "God," the Bishop exclaims, "made this numerous peo-ple very simple, without trickery or malice, most obedient and faithful to their natural lords, and to the Spaniards, whom they serve; most humble, most patient, very peaceful and manageable, without quarrels, strife, bitterness or hate, none desiring vengeance. They are also a very delicate and tender folk, of slender build, and cannot stand much work, and often die of whatever sicknesses they have; so that even our own princes and lords, cared for with all con-veniences, luxuries and delights, are not more delicate than these people who possess little, and who do not desire many worldly goods; nor are they proud, ambitious, or covetous.... They have a very clear and lively understanding, being docile and able to receive all good doctrine, quite fitted to understand our holy Catholic faith, and to be instructed in good and virtuous habits, having less hin-drances in the way of doing this than any other people in the world.... Certainly these people would be the happiest in the world if only they knew God."

But the Spaniards have dealt with these poor souls most mon-strously. "Among these tender lambs, so highly qualified and endowed by their Lord and Creator, the Spaniards have made entrance, like wolves, lions and tigers made cruel by long fasting, and have done nothing in those parts for forty years but cut them in pieces, slaughter them, torture them, afflict them, torment them and destroy them by strange sorts of cruelty never before seen or read or heard ... so that of the three million and more souls who inhabited the Island of Hispaniola ... there are now no more than two hundred natives of that land." The pleasant impression made upon the Indians by the comparative clemency of Columbus has been completely eradicated. "The Indians began to see that these men could not have come from heaven."

The Apostle to the Indians is terribly in earnest. He knows the Indians, and loves them as a father loves his children. He does not claim perfection for them, but he recognizes them as perfectible. He does not assert their superiority to the Spaniards, but his indignation against his countrymen contains the germs of such an assertion.

English views of savage life tend to be less highly colored and enthusiastic than those of the Spanish and French. But though it seems probable that the Noble Savage is chiefly a product of Latin minds, Professor Chinard slightly underestimates the extent to which English explorers gave support to the cult of the Indian.

There are, for example, decidedly sympathetic passages in the *Voyage of Sir Francis Drake from New Spain to the North-west of California*. This celebrated voyage was begun in 1577. The narrator reports that the savages—here natives of Brazil—go stark naked, but he does not philosophize upon this observation. The "naturals" seem to be a civil and gentle folk: "Our general went to prayer ... at which exercise they were attentive and seemed greatly to be affected with it." The savages, indeed, worship the whites as gods, at first making sacrifice to them by tearing their own flesh, and when this is frowned upon by the voyagers, bringing offerings of fruit. The savage king and his people crown Drake with flowers, "with one consent and with great reverence, joyfully singing a song." They wish the English to remain with them for ever. "Our departure seemed so grievous to them, that their joy was turned into sorrow." Incidents such as these are ready-made for literary treatment.

Strenuous efforts were being made to "boom" Virginia as a field of colonization. This may partly account for the enthusiasm of Philip Amadas and Arthur Barlow in their *First Voyage Made to the Coast of Virginia*. These gentlemen find the natives fearless and trustful. They are "a handsome and goodly people, and in their behavior as mannerly and civil as any in Europe." Later it is reported: "We found the people most gentle, loving and faithful, void of all guile and treason, and such as live after the manner of the golden age." This comparison with the Golden Age is particularly interesting. When men began to think of the American Indian in terms of traditional literary formulas, they were well on the way toward the formation of the Noble Savage idea.

A very influential account was doubtless Raleigh's *Discourse of the large, rich and beautiful Empire of Guiana*. The portions of this account which are of interest to us deal with various tribes along the Orinoco River—a region which is the habitat of the Noble Savage at his noblest and most savage.

Raleigh's opinion of the natives is consistently favorable. Of one tribe he says, "These Tivitivas are a very goodly people and very valiant, and have the most manly speech and most deliberate that ever I heard, of what nation soever." This tribe relies for sustenance entirely on the bounty of nature. "They never eat of anything that is set or sowen: and as at home they use neither planting nor other manurance, so when they come abroad, they refuse to feed of aught, but of that which nature without labour bringeth forth."

Raleigh agrees with many other voyagers in ascribing rare physical beauty to the savages. Of a Cacique's wife he writes: "In all my life I have seldome seene a better favoured woman. She was of good stature, with blacke eyes, fat of body, of an excellent countenance, her hair almost as long as herself, tied up againe in prettie knots.... I have seene a lady in England as like to her, as but for the colour, I would have sworne might have been the same." Praise from Sir Hubert!

The following is a portion of an account of an interview with a venerable chief: "I asked what nations those were which inhabited on the farther side of those mountains.... He answered with a great sigh (as a man which had inward feeling of the losse of his countrie and libertie, especially for that his eldest son was slain in a battell on that side of the mountains, whom he most entirely loved) that hee remembered in his father's lifetime, etc., etc.... After hee had answered thus farre he desired leave to depart, saying that he had farre to goe, that he was olde, and weake, and was every day called for by death, which was also his owne phrase.... This Topiawari is helde for the prowdest and wisest of all the Orenoqueponi, and soe he behaved himselfe towards mee in all his answers at my returne, as I marvelled to find a man of that gravitie and judgement, and of soe good discourse, that had no helpe of learning nor breede."

This sketch of the old Cacique is executed with a significant relish. Quite plainly, the savage has become literary material; his type is becoming fixed; he already begins to collect the accretions of tradition. Just as he is, Topiawari is ready to step into an exotic tale. He is the prototype of Chactas and Chingachgook.

The effect on English writers of such accounts as those we have been examining is shown in Michael Drayton's poem, *To the Virginian Voyage*:

> And cheerfully at sea,
> Success you still entice,
>   To get the pearl and gold,
>   And ours to hold
> Virginia,
> Earth's only paradise.
>
> Where nature hath in store
> Fowl, venison, and fish,
>   And the fruitful'st soil,
>   Without your toil,
> Three harvests more,
> All greater than you wish.
>
> To whom the Golden Age
> Still nature's laws doth give,
>   No other cares attend,
>   But them to defend
> From winter's rage,
> That long there doth not live.

Virginia reminds the poet both of the Earthly Paradise and the Golden Age; and the second stanza quoted brings an unconsciously ironical reminder of the *Land of Cockayne*. Here we see that fusion of contemporary observation with old tradition on which the Noble Savage idea depends.

Discourse on the Origins of Inequality (1754)

Jean-Jacques Rousseau

Man, whatever Country you may come from, whatever your opinions may be, listen: here is your history as I believed it to read, not in the Books of your Fellow-men, who are liars, but in Nature, which never lies. Everything that comes from Nature will be true; there will be nothing false except what I have involuntarily put in of my own. The times of which I am going to speak are very far off: how you have changed from what you were! It is, so to speak, the life of your species that I am going to describe to you according to the qualities you received, which your education and habits have been able to corrupt but have not been able to destroy. There is, I feel, an age at which the individual man would want to stop: you will seek the age at which you would desire your Species had stopped. Discontented with your present state for reasons that foretell even greater discontents for your unhappy Posterity, perhaps you would want to be able to go backward in time. This sentiment must be the Eulogy of your first ancestors, the criticism of your contemporaries, and the dread of those who will have the unhappiness to live after you....

Stripping this Being, so constituted, of all the supernatural gifts he could have received and of all the artificial faculties he could only have acquired by long progress—considering him, in a word, as he must have come from the hands of Nature—I see an animal less strong than some, less agile than others, but all things considered, the most advantageously organized of all. I see him satisfying his hunger under an oak, quenching his thirst at the first Stream, finding his bed at the foot of the same tree that furnished his meal; and therewith his needs are satisfied.

The Earth, abandoned to its natural fertility and covered by immense forests never mutilated by the Axe, offers at every step Storehouses and shelters to animals of all species. Men, dispersed among the animals, observe and imitate their industry, and thereby develop in themselves the instinct of the Beasts; with the advantage that whereas each species has only its own proper instinct, man—perhaps having none that belongs to him—appropriates them all to himself, feeds himself equally well with most of the diverse foods which the other animals share, and consequently finds his subsistence more easily than any of them can....

The savage man's body being the only implement he knows, he employs it for various uses of which, through lack of training, our bodies are incapable; our industry deprives us of the strength and agility that necessity obliges him to acquire. If he had an axe, would his wrist break such strong branches? If he had a sling, would he throw a stone so hard? If he had a ladder, would he climb a tree so nimbly? If he had a Horse, would he run so fast? Give Civilized man time to assemble all his machines around him and there can be no doubt that he will easily overcome Savage man. But if you want to see an even more unequal fight, put them, naked and disarmed, face to face, and you will soon recognize the advantage of constantly having all of one's strength at one's disposal, of always being ready for any event, and of always carrying oneself, so to speak, entirely with one.

Hobbes claims that man is naturally intrepid and seeks only to attack and fight. An illustrious Philosopher thinks, on the contrary, and Cumberland and Pufendorf also affirm, that nothing is so timid as man in the state of Nature, and that he is always trembling and ready to flee at the slightest noise he hears, at the slightest movement he perceives. That may be so with respect to objects he does not know; and I do not doubt that he is frightened by all the new Spectacles that present themselves to him every time he can neither discern the Physical good and evil to be expected nor compare his strength with the dangers he must run: rare circumstances in the state of Nature, where all things move in such a uniform manner, and where the face of the Earth is not subject to those brusque and continual changes caused by the passions and inconstancy of united Peoples. But Savage man, living dispersed among the animals and early finding himself in a position to measure himself against them, soon makes the comparison; and sensing that he surpasses them in skill more than they surpass him in strength, he learns not to fear them any more. Pit a bear or a wolf against a Savage who is robust, agile, courageous, as they all are, armed with stones and a good stick, and you will see that the danger will be reciprocal at the very least, and that after several similar experiences wild Beasts, which do not like to attack each other, will hardly attack man willingly, having found him to be just as wild as they.

With regard to animals that actually have more strength than man has skill, he is in the position of the other weaker species, which nevertheless subsist. But man has the advantage that, no less adept at running than they and finding almost certain refuge in trees, he always has the option of accepting or leaving the encounter and the choice of flight or combat. Let us add that it does not appear that any animal naturally makes war upon man except in case of self-defense or extreme hunger, or gives evidence of those violent antipathies toward him that seem to announce that one species is destined by Nature to serve as food for the other.

These are, without doubt, the reasons why Negroes and Savages trouble themselves so little about the wild beasts they may encounter in the woods. In this respect the Caribs of Venezuela, among others, live in the most profound security and without the slightest inconvenience. Although they go nearly naked, says François Corréal, they nevertheless expose themselves boldly in the woods armed only with bow and arrow, but no one has ever heard that any of them were devoured by beasts.

Other more formidable enemies, against which man does not have the same means of defense, are natural infirmities: infancy, old age, and illnesses of all kinds, sad signs of our weakness, of which the first two are common to all animals and the last belongs principally to man living in Society. I even observe on the subject of Infancy that the Mother, since she carries her child with her everywhere, can nourish it with more facility than the females of several animals, which are forced to come and go incessantly with great fatigue, in one direction to seek their food and in the other to suckle or nourish their young. It is true that if the woman should die, the child greatly risks dying with her; but this danger is common to a hundred other species, whose young are for a long time unable to go and seek their nourishment themselves. And if Infancy is longer among us, so also is life; everything remains approximately equal in this respect, although there are, concerning the duration of the first age and the number of young, other rules which are not within my Subject. Among the Aged, who act and perspire little, the need for food diminishes with the faculty of providing for it; and since Savage life keeps gout and rheumatism away from them and since old age is, of all ills, the one that human assistance can least relieve, they finally die without it being perceived that they cease to be, and almost without perceiving it themselves.

With regard to illnesses, I shall not repeat the vain and false declamations against Medicine made by most People in good health; rather, I shall ask whether there is any solid observation from which one might conclude that in Countries where this art is most neglected, the average life of man is shorter than in those where it is cultivated with the greatest care. And how could that be, if we give ourselves more ills than Medicine can furnish Remedies? The extreme inequality in our way of life: excess of idleness in some, excess of labor in others; the ease of stimulating and satisfying our appetites and our sensuality; the overly refined foods of the rich, which nourish them with binding juices and overwhelm them with indigestion; the bad food of the Poor, which they do not even have most of the time, so that their want inclines them to overburden their stomachs greedily when the occasion permits; late nights, excesses of all kinds, immoderate ecstasies of all the Passions, fatigues and exhaustion of Mind; numberless sorrows and afflictions which are felt in all conditions and by which souls are perpetually tormented: these are the fatal proofs that most of our ills are our own work, and that we would have avoided almost all of them by preserving the simple, uniform, and solitary way of life prescribed to us by Nature. If she destined us to be healthy, I almost dare affirm that the state of reflection is a state contrary to Nature and that the man who meditates is a depraved animal. When one thinks of the good constitution of Savages, at least of those whom we have not ruined with our strong liquors; when one learns that they know almost no illnesses except wounds and old age, one is strongly inclined to believe that the history of human illnesses could easily be written by following that of civil Societies. This at least is the opinion of Plato, who judges, from certain Remedies used or approved by Podalirius and Machaon at the siege of Troy, that various illnesses that should have been caused by those remedies were not yet known at that time among men; and Paracelsus reports that the diet, so necessary today, was invented only by Hippocrates.

With so few sources of illness, man in the state of Nature hardly has need of remedies, still less of Doctors. In this respect the human species is not in any worse condition than all the others; and it is easy to learn from Hunters whether in their chases they find many sick animals. They find many that have received extensive but very well healed wounds, that have had bones and even limbs broken and set again with no other Surgeon than time, no other regimen than their ordinary life, and that are no less perfectly cured for not having been tormented with incisions, poisoned with Drugs, or weakened with fasting. Finally, however useful well-administered medicine may be among us, it is still certain that if a sick Savage abandoned to himself has nothing to hope for except from Nature, in return he has nothing to fear except from his illness, which often renders his situation preferable to ours.

Let us therefore take care not to confuse Savage man with the men we have before our eyes. Nature treats all the animals abandoned to its care with a partiality that seems to show how jealous it is of this right. The Horse, the Cat, the Bull, even the Ass, are mostly taller, and all have a more robust constitution, more vigor, more strength and courage in the forest than in our houses. They lose half of these advantages in becoming Domesticated, and it might be said that all our cares to treat and feed these animals well end only in their degeneration. It is the same even for man. In becoming sociable and a Slave he becomes weak, fearful, servile; and his soft and effeminate way of life completes the enervation of both his strength and his courage. Let us add that between Savage and Domesticated conditions the difference from man to man must be still greater than that from beast to beast; for animal and man having been treated equally by Nature, all the commodities of which man gives himself more than the animals he tames are so many particular causes that make him degenerate more noticeably....

The example of Savages, who have almost all been found at this point, seems to confirm that the human Race was made to remain in it, the state of Nature, always; that this state is the veritable youth of the World; and that all subsequent progress has been in appearance so many steps toward the perfection of the individual, and in fact toward the decrepitude of the species.

As long as men were content with their rustic huts, as long as they were limited to sewing their clothing of skins with thorn or fish bones, adorning themselves with feathers and shells, painting their bodies with various colors, perfecting or embellishing their bows and arrows, carving with sharp stones a few fishing Canoes or a few crude Musical instruments; in a word, as long as they applied themselves only to tasks that a single person could do and to arts that did not require the cooperation of several hands, they lived free, healthy, good, and happy insofar as they could be according to their Nature, and they continued to enjoy among themselves the sweetness of independent intercourse. But from the moment one man needed the help of another, as soon as they observed that it was useful for a single person to have provisions for two, equality disappeared, property was introduced, labor became necessary; and vast forests were changed into smiling Fields which had to be watered with the sweat of men, and in which slavery and misery were soon seen to germinate and grow with the crops.

Metallurgy and agriculture were the two arts whose invention produced this great revolution. For the Poet it is gold and silver, but for the Philosopher it is iron and wheat which have Civilized men and ruined the human Race.

"Excursions" (1863)

Henry David Thoreau

I believe in the forest, and in the meadow, and in the night in which the corn grows. We require an infusion of hemlock spruce or arbor-vitae in our tea. There is a difference between eating and drinking for strength and from mere gluttony. The Hottentots eagerly devour the marrow of the koodoo and other antelopes raw, as a matter of course. Some of our northern Indians eat raw the marrow of the Arctic reindeer, as well as various other parts, including the summits of the antlers, as long as they are soft. And herein, perchance, they have stolen a march on the cooks of Paris. They get what usually goes to feed the fire. This is probably better than stall-fed beef and slaughter-house pork to make a man of. Give me a wildness whose glance no civilization can endure,—as if we lived on the marrow of koodoos devoured raw.

There are some intervals which border the strain of the wood thrush, to which I would migrate,—wild lands where no settler has squatted; to which, methinks, I am already acclimated.

The African hunter Cumming tells us that the skin of the eland, as well as that of most other antelopes just killed, emits the most delicious perfume of trees and grass. I would have every man so much like a wild antelope, so much a part and parcel of nature, that his very person should thus sweetly advertise our senses of his presence, and remind us of those parts of nature which he most haunts. I feel no disposition to be satirical, when the trapper's coat emits the odor of musquash even; it is a sweeter scent to me than that which commonly exhales from the merchant's or the scholar's garments. When I go into their wardrobes and handle their vestments, I am reminded of no grassy plains and flowery meads which they have frequented, but of dusty merchants' exchanges and libraries rather.

A tanned skin is something more than respectable, and perhaps olive is a fitter color than white for a man—a denizen of the woods. "The pale white man!" I do not wonder that the African pitied him. Darwin the naturalist says, "A white man bathing by the side of a Tahitian was like a plant bleached by the gardener's art, compared with a fine, dark green one, growing vigorously in the open fields."

Ben Jonson exclaims,—
How near to good is what is fair!
So I would say—
How near to good is what is wild!

Life consists with wildness. The most alive is the wildest. Not yet subdued to man, its presence refreshes him. One who pressed forward incessantly and never rested from his labors, who grew fast and made infinite demands on life, would always find himself in a new country or wilderness, and surrounded by the raw material of life. He would be climbing over the prostrate stems of primitive forest-trees.

Against His-story, Against Leviathan! (1983)

Fredy Perlman

The managers of Gulag's islands tell us that the swimmers, crawlers, walkers and fliers spent their lives working in order to eat.

These managers are broadcasting their news too soon. The varied beings haven't all been exterminated yet. You, reader, have only to mingle with them, or just watch them from a distance, to see that their waking lives are filled with dances, games and feasts. Even the hunt, the stalking and feigning and leaping, is not what we call Work, but what we call Fun. The only beings who work are the inmates of Gulag's islands, the zeks.

The zeks' ancestors did less work than a corporation owner. They didn't know what work was. They lived in a condition J.J. Rousseau called "the state of nature." Rousseau's term should be brought back into common use. It grates on the nerves of those who, in R. Vaneigem's words, carry cadavers in their mouths. It makes the armor visible. Say "the state of nature" and you'll see the cadavers peer out.

Insist that "freedom" and "the state of nature" are synonyms, and the cadavers will try to bite you. The tame, the domesticated, try to monopolize the word freedom; they'd like to apply it to their own condition. They apply the word "wild" to the free. But it is another public secret that the tame, the domesticated, occasionally become wild but are never free so long as they remain in their pens.

Even the common dictionary keeps this secret only half hidden. It begins by saying that free means citizen! But then it says, "Free: a) not determined by anything beyond its own nature or being; b) determined by the choice of the actor or by his wishes...."

The secret is out. Birds are free until people cage them. The Biosphere, Mother Earth herself, is free when she moistens herself, when she sprawls in the sun and lets her skin erupt with varicolored hair teeming with crawlers and fliers. She is not determined by anything beyond her own nature or being until another sphere of equal magnitude crashes into her, or until a cadaverous beast cuts into her skin and rends her bowels.

Trees, fish and insects are free as they grow from seed to maturity, each realizing its own potential, its wish—until the insect's freedom is curtailed by the bird's. The eaten insect has made a gift of its freedom to the bird's freedom. The bird, in its turn, drops and manures the seed of the insect's favorite plant, enhancing the freedom of the insect's heirs.

The state of nature is a community of freedoms.

Such was the environment of the first human communities, and such it remained for thousands of generations.

Modern anthropologists who carry Gulag in their brains reduce such human communities to the motions that look most like work, and give the name Gatherers to people who pick and sometimes store their favorite foods. A bank clerk would call such communities Savings Banks!

The zeks on a coffee plantation in Guatemala are Gatherers, and the anthropologist is a Savings Bank. Their free ancestors had more important things to do.

The !Kung people miraculously survived as a community of free human beings into our own exterminating age. R.E. Leakey observed them in their lush African forest homeland. They cultivated nothing except themselves. They made themselves what they wished to be. They were not determined by anything beyond their own being—not by alarm clocks, not by debts, not by orders from superiors. They feasted and celebrated and played, full-time, except when they slept. They shared everything with their communities: food, experiences, visions, songs. Great personal satisfaction, deep inner joy, came from the sharing.

(In today's world, wolves still experience the joys that come from sharing. Maybe that's why governments pay bounties to the killers of wolves.)

S. Diamond observed other free human beings who survived into our age, also in Africa. He could see that they did no work, but he couldn't quite bring himself to say it in English. Instead, he said they made no distinction between work and play. Does Diamond mean that the activity of the free people can be seen as work one moment, as play another, depending on how the anthropologist feels? Does he mean they didn't know if their activity was work or play? Does he mean we, you and I, Diamond's armored contemporaries, cannot distinguish their work from their play?

If the !Kung visited our offices and factories, they might think we're playing. Why else would we be there?

I think Diamond meant to say something more profound. A time-and-motion engineer watching a bear near a berry patch would not know when to punch his clock. Does the bear start working when he walks to the berry patch, when he picks the berry, when he opens his jaws? If the engineer has half a brain he might say the bear makes no distinction between work and play. If the engineer has an imagination he might say that the bear experiences joy from the moment the berries turn deep red, and that none of the bear's motions are work.

Leakey and others suggest that the general progenitors of human beings, our earliest grandmothers, originated in lush African forests, somewhere near the homeland of the !Kung. The conservative majority, profoundly satisfied with nature's unstinting generosity, happy in their accomplishments, at peace with themselves and the world, had no reason to leave their home. They stayed.

A restless minority went wandering. Perhaps they followed their dreams. Perhaps their favorite pond dried up. Perhaps their favorite animals wandered away. These people were very fond of animals; they knew the animals as cousins.

The wanderers are said to have walked to every woodland, plain and lakeshore of Eurasia. They walked or floated to almost every island. They walked across the land bridge near the northern land of ice to the southernmost tip of the double continent which would be called America.

The wanderers went to hot lands and cold, to lands with much rain and lands with little. Perhaps some felt nostalgia for the warm home they left. If so, the presence of their favorite animals, their cousins, compensated for their loss. We can still see the homage some of them gave to these animals on cave walls of Altamira, on rocks in Abrigo del Sol in the Amazon Valley.

Some of the women learned from birds and winds to scatter seeds. Some of the men learned from wolves and eagles to hunt.

But none of them ever worked. And everyone knows it. The armored Christians who later "discovered" these communities knew that these people did no work, and this knowledge grated on Christian nerves, it rankled, it caused cadavers to peep out. The Christians spoke of women who did "lurid dances" in their fields instead of confining themselves to chores; they said hunters did a lot of devilish "hocus pocus" before actually drawing the bowstring.

These Christians, early time-and-motion engineers, couldn't tell when play ended and work began. Long familiar with the chores of zeks, the Christians were repelled by the lurid and devilish heathen who pretended that the Curse of Labor had not fallen on them. The Christians put a quick end to the "hocus pocus" and the dances, and saw to it that none could fail to distinguish work from play.

Our ancestors—I'll borrow Turner's term and call them the Possessed—had more important things to do than to struggle to survive. They loved nature and nature reciprocated their love. Wherever they were they found affluence, as Marshall Sahlins shows in his *Stone Age Economics*. Pierre Clastres' *La société contre l'état* insists that the struggle for subsistence is not verifiable among any of the Possessed; it is verifiable among the Dispossessed in the pits and on the margins of progressive industrialization. Leslie White, after a sweeping review of reports from distant places and ages, a view of "Primitive culture as a whole," concludes that "there's enough to eat for a richness of life rare among the 'civilized.'" I wouldn't use the word Primitive to refer to people with a richness of life. I would use the word Primitive to refer to myself and my contemporaries, with our progressive poverty of life.

Primitive Man and His Food (1952)

Arnold DeVries

The defective state of modern man has had its effects upon medicine and the very study of disease. Dr. E.A. Hooton, the distinguished physical anthropologist of Harvard, has remarked that "it is a very myopic medical science which works backward from the morgue rather than forward from the cradle." Yet this is exactly what the customary procedure of medicine has been. The reasons have been somewhat of necessity, it is to be admitted, for one can scarcely study health when the adequate controls are not present. In civilization one studies civilized people, and the frequency of the forms of degeneration which are found then determine what we consider normal and abnormal. As a result, conditions which generally form no part of undomesticated animal life are regarded as normal and necessary for the human species. So long has disease been studied that the physician often has little concept as to what health actually is. We live in a world of pathology, deformity and virtual physical monstrosity, which has so colored our thinking that we cannot visualize the nature of health and the conditions necessary for its presence.

The question should then logically arise: why not leave civilization and study physical conditions in the primitive world? If perfect physical specimens could here be found, the study could be constructive and progressive, giving suggestions, perhaps, as to the conditions which permitted or induced a state of physical excellence to exist. We might then find out what man is like, biologically speaking, when he does not need a doctor, which might also indicate what he should be like when the doctor has finished with him.

Fortunately the idea has not been entirely neglected. Primitive races were carefully observed and described by many early voyagers and explorers who found them in their most simple and natural state. Primitive life has also very carefully been observed and studied with the object of understanding social, moral or religious conditions, in which, however, incidental observations were made too with respect to the physical condition of the people, and the living habits which might affect that condition. Others, in modern life, have studied the savages with the specific object of determining their physical state of health, and the mode of living which is associated therewith.

The results of such work have been very significant, but regarding medicine and nutrition in actual practice, they have been almost entirely neglected. The common view that primitive man is generally short lived and subject to many diseases is often held by physician as well as layman, and the general lack of sanitation, modern treatment, surgery and drugs in the primitive world is thought to prevent maintenance of health at a high physical level. For the average nutritionist it is quite natural to feel that any race not having access to the wide variety of foods which modern agriculture and transportation now permit could not be in good health. These assumptions have helped to determine existing therapeutic methods, and they have largely prevented serious consideration that might be based upon factual data.

But the facts are known, and these comprise a very interesting and important story. They indicate that, when living under near-isolated conditions, apart from civilization and without access to the foods of civilization, primitive man lives in much better physical condition than does the usual member of civilized society. When his own nutrition is adequate and complete, as it often is, he maintains complete immunity to dental caries. His teeth are white and sparkling, with neither brushing nor cleansing agents used, and the dental arch is broad, with the teeth formed in perfect alignment.

The facial and body development is also good. The face is finely formed, well-set and broad. The body is free from deformity and proportioned as beauty and symmetry would indicate desirable. The respective members of the racial group reproduce in homogeneity from one generation to the next. There are few deviations from the standard anthropological prototype. One individual resembles the other in facial form, looking much like sisters or brothers, with the chief differences in appearance being in size.

Reproductive efficiency is such as to permit parturition with no difficulty and little or no pain. There are no prenatal deformities. Resistance to infectious disease is high, few individuals being sick, and these usually rapidly recovering. The degenerative diseases are rare, even in advanced life, some of them being completely unknown and unheard of by the primitive. Mental complaints are equally rare, and the state of happiness and contentment is one scarcely known by civilized man. The duration of life is long, the people being yet strong and vigorous as they pass the proverbial three score and ten mark, and living in many cases beyond a century.

These are the characteristics of the finest and most healthful primitive races, who live under the most ideal climatic and nutritional conditions. Primitive races less favored by environment are less successful in meeting weakness and disease, but even the poorest of these have better teeth and skeletal development than civilized man, and they usually present other physical advantages as well.

The experience of primitive man has therefore been one of great importance. We note that people living today, under the culture and environment of the Stone Age, have not only equalled but far surpassed civilized man in strength, physical development and immunity to disease. The mere existence of this fact poses an important question to modern medicine and should arouse serious thought and consideration.

Of equal significance is the fact that the good health of the primitive has been possible only under conditions of relative isolation. As soon as his contact with civilization is sufficient to alter his dietary habits, he succumbs to disease very readily and loses all of the unique immunity of the past. The teeth decay; facial form ceases to be uniform; deformities become common; reproductive efficiency is lowered; mental deficiency develops; and the duration of life is sharply lowered.

It would hence appear that the nutritional habits of primitive man are responsible for his state of health. So long as the native foods remain in use, there are no important physical changes, and the bacterial scourges are absent, even though a complete lack of sanitation would indicate that pathogenic bacteria might be present. With a displacement of native foods for those of modern commerce the situation changes completely, and the finest sanitation that the white man can provide, together with the best in medical services, is of no avail in preventing the epidemics which take thousands of lives. Among scientists who have studied at first hand both the physical condition and food of many primitive races, the close relationship between the two has been clearly recognized.

"The Original Affluent Society" (1968)

Marshall Sahlins

S f economics is the dismal science, the study of hunting-gathering economies must be its most advanced branch. Almost totally committed to the argument that life was hard in the Paleolithic, our textbooks compete to convey a sense of impending doom, leaving the student to wonder not only how hunters managed to make a living, but whether, after all, this was living? The specter of starvation stalks the stalker in these pages. His technical incompetence is said to enjoin continuous work just to survive, leaving him without respite from the food quest and without the leisure to "build culture." Even so, for his efforts he pulls the lowest grades in thermo-dynamics—less energy harnessed per capita per year than any other mode of production. And in treatises on economic development, he is condemned to play the role of bad example, the so-called "subsistence economy."

It will be extremely difficult to correct this traditional wisdom. Perhaps then we should phrase the necessary revisions in the most shocking terms possible: that this was, when you come to think of it, the original affluent society. By common understanding an affluent society is one in which all the people's wants are easily satisfied; and though we are pleased to consider this happy condition the unique achievement of industrial civilization, a better case can be made for hunters and gatherers, even many of the marginal ones spared to ethnography. For wants are "easily satisfied," either by producing much or desiring little, and there are, accordingly, two possible roads to affluence. The Galbraithean course makes assumptions peculiarly appropriate to market economies, that man's wants are great, not to say infinite, whereas his means are limited, although improvable. Thus the gap between means and ends can eventually be narrowed by industrial productivity, at least to the extent that "urgent" goods became abundant. But there is also a Zen solution to scarcity and affluence, beginning from premises opposite from our own, that human material ends are few and finite and technical means unchanging but on the whole adequate.

Adopting the Zen strategy, a people can enjoy an unparalleled material plenty, though perhaps only a low standard of living. That I think describes the hunters.

The traditional dismal view of the hunter's fix is pre-anthropological. It goes back to the time Adam Smith was writing, and maybe to a time before anyone was writing. But anthropology, especially evolutionary anthropology, found it congenial, even necessary theoretically, to adopt the same tone of reproach. Archeologists and ethnologists had become Neolithic revolutionaries, and in their enthusiasm for the revolution found serious shortcomings in the Old (Stone Age) Regime. Scholars extolled a Neolithic Great Leap Forward. Some spoke of a changeover from human effort to domesticated energy sources, as if people had been liberated by a new labor-saving device, although in fact the basic power resources remained exactly the same, plants and animals, the development occurring rather in techniques of appropriation (i.e., domestication. Moreover, archeological research was beginning to suggest that the decisive gains came in stability of settlement and gross economic product, rather than productivity of labor.)

But evolutionary theory is not entirely to blame. The larger economic context in which it operates, "as if by an invisible hand," promotes the same dim conclusions about the hunting life. Scarcity is the peculiar obsession of a business economy, the calculable condition of all who participate in it. The market makes freely available a dazzling array of products all these "good things" within a man's reach—but never his grasp, for one never has enough to buy everything. To exist in a market economy is to live out a double tragedy, beginning in inadequacy and ending in deprivation. All economic activity starts from a position of shortage: whether as producer, consumer, or seller of labor, one's resources are insufficient to the possible uses and satisfactions. So one comes to a conclusion— "you pays your money and you takes your choice." But then, every acquisition is simultaneously a deprivation, for every purchase of something is a denial of something else that could have been had instead. (The point is that if you buy one kind of automobile, say a Plymouth fastback, you cannot also have a Ford Mustang and I judge from the TV commercials that the deprivation involved is more than material.) Inadequacy is the judgment decreed by our economy, and thus the axiom of our economics: the application of scarce means against alternate ends. We stand sentenced to life at hard labor. It is from this anxious vantage that we look back on the hunter. But if modern man, with all his technical advantages, still hasn't got the wherewithal, what chance has this naked savage with his puny bow and arrow? Having equipped the hunter with bourgeois impulses and Paleolithic tools, we judge his situation hopeless in advance.

Scarcity is not an intrinsic property of technical means. It is a relation between means and ends. We might entertain the empirical possibility that hunters are in business for their health, a finite objective, and bow and arrow are adequate to that end. A fair case can be made that hunters often work much less than we do, and rather than a grind the food quest is intermittent, leisure is abundant, and there is more sleep in the daytime per capita than in any other conditions of society. (Perhaps certain traditional formulae are better inverted: the amount of work per capita increases with the evolution of culture and the amount of leisure per capita decreases.) Moreover, hunters seem neither harassed nor anxious. A certain confidence, at least in many cases, attends their economic attitudes and decisions. The way they dispose of food on hand, for example—as if they had it made.

This is the case even among many present marginal hunters—who hardly constitute a fair test of Paleolithic economy but something of a supreme test. Considering the poverty in which hunter and gatherers live in theory, it comes as a surprise that Bushmen who live in the Kalahari enjoy "a kind of material plenty" (Marshall, 1961, p. 243). Marshall is speaking of non-subsistence production; in this context her explication seems applicable beyond the Bushmen. She draws attention to the technical simplicity of the non-subsistence sector: the simple and readily available raw materials, skills, and tools. But most important, wants are restricted: a few people are happy to consider few things their good fortune. The restraint is imposed by nomadism. Of the hunter, it is truly said that this wealth is a burden (at least for his wife). Goods and mobility are therefore soon brought into contradiction, and to take liberties with a line of Lattimore's, the pure nomad remains a poor nomad. It is only consistent with their mobility, as many accounts directly say, that among hunters needs are limited, avarice inhibited, and—Warner (1937 [1958], p. 137) makes this very clear for the Murngin—portability is a main value in the economic scheme of things.

A similar case of affluence without abundance can be made for the subsistence sector. McCarthy and McArthur's time-motion study in Arnhem Land (1960) indicates the food quest is episodic and discontinuous, and per capita commitment to it averages less than four hours a day. The amount of daytime sleep and rest is unconscionable: clearly, the aborigines fail to "build culture" not from lack of time but from idle hands. McCarthy and McArthur also suggest that the people are working under capacity—they might have easily procured more food; that they are able to support unproductive adults—who may, however, do some craft work; and that getting food was not strenuous or exhausting. The Arnhem Land study, made under artificial conditions and based only on short-run observations, is plainly inconclusive in itself. Nevertheless, the Arnhem Land data are echoed in reports of other Australians and other hunters. Two famous explorers of the earlier nineteenth century made estimates of the same magnitude for the aborigines' subsistence activities: two to four hours a day (Eyre, 1845, 2, pp. 252, 255; Grey, 1841, 2, pp. 261–63). Slash-and-burn agriculture, incidentally, may be more labor-intensive: Conklin, for example, figures that 1,200 man hours per adult per year are given among the Hanunóo simply to agriculture (Conklin, 1957, p. 151: this figure excludes other food-connected activities, whereas the Australian data include time spent in the preparation of food as well as its acquisition). The Arnhem Landers' punctuation of steady work with sustained idleness is also widely attested in Australia and beyond. In Lee's paper he reported that productive members of !Kung Bushman camps spend two to three days per week in subsistence. We have heard similar comments in other papers at the symposium. Hadza women were said to work two hours per day on the average in gathering food, and one concludes from James Woodburn's excellent film that Hadza men are much more preoccupied with games of chance than chances of game (Woodburn and Hudson, 1966).

In addition, evidence on hunter-gatherers' economic attitudes and decisions should be brought to bear. Harassment is not implied in the descriptions of their nonchalant movements from camp to camp, nor indeed is the familiar condemnations of their laziness. A certain issue is posed by exasperated comments on the prodigality of hunters, their inclination to make a feast of everything on hand; as if, one Jesuit said of the Montagnais, "the game they were to hunt was shut up in a stable" (Le Jeune's *Relation* of 1634, in Kenton, 1927, I, p. 182). "Not the slightest thought of, or care for, what the morrow may bring forth," wrote Spencer and Gillen (1899, p. 53). Two interpretations of this supposed lack of foresight are possible:

either they are fools, or they are not worried—that is, as far as they are concerned, the morrow will bring more of the same. Rather than anxiety, it would seem the hunters have a confidence born of affluence, of a condition in which all the people's wants (such as they are) are generally easily satisfied. This confidence does not desert them during hardship. It can carry them laughing through periods that would try even a Jesuit's soul, and worry him so that—as the Indians warn—he could become sick:

"I saw them [the Montagnais] in their hardships and their labors, suffer with cheerfulness.... I found myself, with them, threatened with great suffering; they said to me, 'We shall be sometimes two days, sometimes three, without eating, for lack of food; take courage, *Chihine*, let thy soul be strong to endure suffering and hardship; keep thyself from being sad, otherwise thou will be sick; see how we do not cease to laugh, although we have little to eat'" (Le Jeune's *Relation* of 1634, in Kenton, 1927, I, p. 129).

Again on another occasion Le Jeune's host said to him: "Do not let thyself be cast down, take courage; when the snow comes, we shall eat" (Le Jeune's *Relation* of 1634, in Kenton, 1927, I, p. 171). Which is something like the philosophy of the Penan of Borneo: "If there is no food today there will be tomorrow"—expressing, according to Needham, "a confidence in the capacity of the environment to support them, and in their own ability to extract their livelihood from it" (1954, p. 230).

"Birds Combat Civilization" (1985)

Lynn Clive

umankind truly was not meant to fly, and birds keep trying to tell us so. As people and their flying machines continue to over-populate the skies, not only do plane-to-plane collisions increase, but bird to plane collisions drastically increase as well, especially since new technology has created sleeker and quieter engines which sneak up on birds and scarcely give them any warning of their approach. Needless to say, it is the birds which must attempt to change their natural flight patterns to avoid fatal collisions.

Seagulls have become a particularly confounding nuisance to airport officials in Michigan. As their natural feeding grounds along the Great Lakes become more and more polluted, they drift inland. Wet runways peppered with worms and grasshoppers provide a perfect new feeding ground for seagulls. Cherry Capital Airport near Traverse City has reported large flocks of seagulls, as many as 150 at a time.

Approximately 1,200 plane-bird collisions occur each year, causing $20–30 million in damage. Such collisions prove fatal for the birds, of course; however, they have also been responsible for many aircraft crashes fatal to human beings. Sixty-two people were killed in 1960 near Boston when a propeller-driven plane sucked in several starlings and lost power.

Birds seem to be waging all-out war against the U.S. Air Force. In 1983, it reported 2,300 bird collisions; and 300 of these each caused more than $1,000 in damage. This past summer in Great Britain, a U.S. Air Force crew was forced to bail out of their F-111 jet when a 12-pound goose smashed into the protective covering on the nose of the jet. The jet, worth $30.9 million, is now quietly at rest on the bottom of the North Sea.

So what does civilized man do to combat the situation? In Traverse City, airport employees run around the airfield chasing gulls away with "cracker shells" fired from shotguns. They play tapes on loudspeakers of the cries of wounded seagulls, and they're considering putting up hawk silhouettes to see if that might do the trick.

Someone has invented something called a "chicken gun" or a "rooster booster" which hurls four-pound chicken carcasses into the windshields of aircraft at speeds over 500 mph to test their strength against bird collisions. These tests are presently taking place on Air Force jets.

BASH (Bird Air Strike Hazard Team) was organized by the U.S. Air Force in 1975 after three F-111 jets were lost due to bird collisions. This team, made up of Air Force biologists, travels to U.S. bases around the world, targeting bird troublespots and trying to come up with innovative ideas (like the rooster booster) to deal with the problem.

Modern industrial-technological civilizations are based on and geared to the destruction of the natural order. They pollute the air and feeding grounds of wildlife; they chase birds from the skies. They construct buildings like the Renaissance Center in Detroit with mirror-like reflective shells which confuse birds and cause them to crash into them.

As our buildings grow taller and as we fly higher and higher, as we overpopulate our skies with our deadly contrivances, we lose sight of our true and now former place on the earth. We myopically look only at tomorrow. We can marvel at the exquisite beauty of a single bird through a pair of binoculars and then, with the same eye, turn and marvel at a newly constructed skyscraper or a supersonic jet—man's artifices which are responsible for killing flocks of such birds.

If anyone were to suggest to the BASH team that the best way to stop bird-plane collisions would be to stop flying altogether, they would, of course, think you insane—or perhaps "bird-brained." But what is so bad about bird brains? If we acknowledge the message our bird cousins are sending us, maybe it isn't such a bad idea after all.

"WILDFLOWERS:
A BOUQUET OF THESES" (1998)

JOHN LANDAU

What I desire is a return to the profundity of experience. I want a society where everyday activity, however mundane, is centered around how incredibly profound everything is. I want that profundity to become so immense that any mediations between us and it become totally unnecessary: we are in the marvel. When I am in that awe, words are so irrelevant, I don't really care if you call my experience "God" or not. All I know is it is the greatest pleasure possible: to hug a tree, to jump up and down at a beautiful sunset, to climb a magnificent hill, to take awe in what surrounds us. I am a hedonist, and I will have these pleasures; neither the religionist nor the atheist shall lock them away from me!

Primal peoples were in touch with this profoundness, and organized their life around it. Religion is a decadent second-hand relic of this original, authentic mode of experiencing, that attempts to blackmail by linking social control and morality with profound experiencings. Primal peoples sought to avoid whatever distracted from this profundity as much as possible. Obsessiveness of any sort could distract from the wholistic goodness of the environment.

Why are we here? *To experience profoundly.*

Our task, therefore, is to rearrange life (society, the economy) such that profundity is *immanent* in everyday life. Spirituality represents the specialization and detachment of profundity from everyday life into a disembodied, disconnected, symbolic realm that becomes compensatory for an everyday life whose immanence is banality. It is obvious that we don't regularly experience wonder, and this is a social-material problem, because the structure of everyday life discourages this. Other societies in history, however, have endeavored to discover what is truly of value in life, and then, and only then to structure everyday life upon those evident values.

We wish to make calculation and obligation islands in a sea of wonder and awe. We wish to make aloneness a positive experience within the context of profound, embodied togetherness.

Western spirituality has perpetuated a separation between the material and spiritual realms, probably because it arose out of a civilization ruled by an out-of-control materialism. The world used to be experienced profoundly; in spiritual terms, the earth used to be inhabited by spirit. Western spirituality abstracted spirit from the world, from the flesh, leaving an enlivened, disembodied spirit and a deadened, barren world. *It is our job to refuse what has been artificially separated*, not through a symbolical gesture, but by existentially redressing the alienations to which we have been subjected.

Human beings have developed over the past two million years various strategies for taking care of what some have called our "needs." Various subsistence strategies have been invented, and our task would be to examine these and choose the strategies which best support an everyday experience of profundity....

We are discussing a life where one gives joy to others through the mere act of being, where exchange of gifts is a way of life, where one's routine has inherent meaning, not because it makes reference to some symbolic system, but because it opens one out onto *kairos*, the profound moment, the experience of ambience, awe.

In order to do this, we must develop a *pace* that is conducive to this, a set of understandings whereby the experience of profundity is a value and for which rests, pauses, and meditations are in order as a part of routine, and a social reality based upon sharing of profound experiences as primary exchange rather than the exchange of money or etiquette.

Our job is to *invent* primal peoples! Through our imagination and what little we know there is no evidence *against* such group movement. We must imagine these primeval peoples, in order to *create* an incredible myth in order to live it, to become it!

Silence was a great future of such times. People gestured towards the world. Experiences of awe, wonder were everyday affairs. Because people lived outside, they had a much greater oxygen content. They lived in a perpetual oxygen bath, which produces highs, heightens the sense of taste and smell, and is very relaxing. Anyone who has camped out in the open air knows this experience.

The energetic connection with the surroundings was immense; an incredible exchange on all levels was constantly taking place. It is within the context of this immenseness that our words, our 'rationality,' our technical pragmatics seem so narrow, so very small. Far from being primitive, these were people enjoying and interested in preserving immenseness. This is no idealism. A concrete experience in nature can demonstrate the incredible power of the outdoors. One may engage in an intense, strenuous experience with others for a few hours (a night hike or somesuch) and then afterwards meander about in total silence, gesturing at most, exploring movement, smells, and impulses. This will give a taste of how rich it all is. This is what we have lost in our narrow obsessiveness with technicality. What Zen practitioners strive for a lifetime for, our ancestors had by birthright. Sure, they didn't know how to make a waterwheel or how to harness electricity; they didn't want to: they had better things to do! It is even remotely conceivable that they did know of these things, in potential form at least, but saw them as trivial to the process of life....

In the silence, all of the chitchat and all of the worries and all of the monuments fade. In the is-ness, what need to leave one's mark? What need to become immortal through art or culture? Disappearance is erasing the record, off track, no trails, no history. One is in the disappearance already. All one needs is to lose track, to stop recording, to turn off the tape machine, to disappear, it's all right.... It's OK to disappear. Do so now. The grass in front of you is all that ever was or will be. It has no memory, no future. Just silence.

So when we know this rich heritage, when we reach into the heart of our being and know that humans very like ourselves lived a good *two million years* in this way of being, we are awed, and the scum at the top of the pond, the curdled milk of history, our obsession with technicality, pours off and we are left with the pure froth of Being.

MINIMA MORALIA:
REFLECTIONS FROM DAMAGED LIFE (1947)

THEODOR ADORNO

oy shop. Hebbel, in a surprising entry in his diary, asks what takes away 'life's magic in later years.' It is because in all the brightly-coloured contorted marionettes, we see the revolving cylinder that sets them in motion, and because for this very reason the captivating variety of life is reduced to wooden monotony. A child seeing the tightrope-walkers singing, the pipers playing, the girls fetching water, the coachmen driving, thinks all this is happening for the joy of doing so; he can't imagine that these people also have to eat and drink, go to bed and get up again. We however, know what is at stake.' Namely, earning a living, which commandeers all those activities as mere means, reduces them to interchangeable, abstract labour-time. The quality of things ceases to be their essence and becomes the accidental appearance of their value. The 'equivalent form' mars all perceptions; what is no longer irradiated by the light of its own self-determination as 'joy in doing,' pales to the eye. Our organs grasp nothing sensuous in isolation, but notice whether a colour, a sound, a movement is there for its own sake or for something else; wearied by a false variety, they steep all in grey, disappointed by the deceptive claim of qualities still to be there at all, while they conform to the purposes of appropriation, indeed largely owe their existence to it alone. Disenchantment with the contemplated world is the sensorium's reaction to its objective role as a 'commodity world.' Only when purified of appropriation would things be colourful and useful at once: under universal compulsion the two cannot be reconciled.

Children, however, are not so much, as Hebbel thought, subject to illusions of 'captivating variety,' as still aware, in their spontaneous perception, of the contradiction between phenomenon and fungibility that the resigned adult no longer sees, and they shun it. Play is their defense. The unerring child is struck by the 'peculiarity of the equivalent form': 'use-value' becomes the form of manifestation, the phenomenal form of its opposite, value.

In his purposeless activity the child, by a subterfuge, sides with use-value against exchange value. Just because he deprives the things with which he plays of their mediated usefulness, he seeks to rescue in them what is benign towards men and not what subserves the exchange relation that equally deforms men and things. The little trucks travel nowhere and the tiny barrels on them are empty; yet they remain true to their destiny by not performing, not participating in the process of abstraction that levels down that destiny, but instead abide as allegories of what they are specifically for. Scattered, it is true, but not ensnared, they wait to see whether society will finally remove the social stigma on them; whether the vital process between men and things, praxis, will cease to be practical. The unreality of games gives notice that reality is not yet real. Unconsciously they rehearse the right life. The relation of children to animals depends entirely on the fact that Utopia goes disguised in the creatures whom Marx even begrudged the surplus value they contribute as workers. In existing without any purpose recognizable to men, animals hold out, as if for expression, their own names, utterly impossible to exchange. This makes them so beloved of children, their contemplation so blissful. I am a rhinoceros, signifies the shape of the rhinoceros. Fairy-tales and operettas know such images, and the ridiculous question ... how do we know that Orion is really called Orion, rises to the stars.

OUR KIND (1989)

MARVIN HARRIS

Can humans exist without some people ruling and others being ruled? The founders of political science did not think so. "I put for a general inclination of mankind, a perpetual and restless desire for power after power, that ceaseth only in death," declared Thomas Hobbes. Because of this innate lust for power, Hobbes thought that life before (or after) the state was a "war of every man against every man"—"solitary, poor, nasty, brutish and short." Was Hobbes right? Do humans have an unquenchable desire for power that, in the absence of a strong ruler, inevitably leads to a war of all against all? To judge from surviving examples of bands and villages, for the greater part of prehistory our kind got along quite well without so much as a paramount chief, let alone the all-powerful English leviathan King and Mortal God, whom Hobbes believed was needed for maintaining law and order among his fractious countrymen.

Modern states with democratic forms of government dispense with hereditary leviathans, but they have not found a way to dispense with inequalities of wealth and power backed up by an enormously complex system of criminal justice. Yet for 30,000 years after takeoff, life went on without kings, queens, prime ministers, presidents, parliaments, congresses, cabinets, governors, mayors, police officers, sheriffs, marshals, generals, lawyers, bailiffs, judges, district attorneys, court clerks, patrol cars, paddy wagons, jails, and penitentiaries. How did our ancestors manage to leave home without them?

Small populations provide part of the answer. With 50 people per band or 150 per village, everybody knew everybody else intimately, so that the bonding of reciprocal exchange could hold people together. People gave with the expectation of taking and took with the expectation of giving. Since chance played a great role in the capture of animals, collection of wild foodstuffs, and the success of rudimentary forms of agriculture, the individuals who had the luck of the catch on one day needed a handout on the next. So the best way for them to provide for their inevitable rainy day was to be generous. As expressed by the anthropologist Richard Gould, "the greater the amount of risk, the greater the extent of sharing." Reciprocity is a small society's bank.

In reciprocal exchange, people do not specify how much or exactly what they expect to get back or when they expect to get it. That would besmirch the quality of the transaction and make it similar to mere barter or to buying and selling. The distinction lingers on in societies dominated by other forms of exchange, even capitalist ones. For we do carry out a give-and-take among close kin and friends that is informal, uncalculating, and imbued with a spirit of generosity. Teenagers do not pay cash for their meals at home or for the use of the family car, wives do not bill their husbands for cooking a meal, and friends give each other birthday gifts and Christmas presents. But much of this is marred by the expectation that our generosity will be acknowledged with expressions of thanks. Where reciprocity really prevails in daily life, etiquette requires that generosity be taken for granted. As Robert Dentan discovered while doing fieldwork among the Semai of central Malaysia, no one ever says "thank you" for the meat received from another hunter. Having struggled all day to lug the carcass of a pig home through the jungle heat, the hunter allows his prize to be cut up into equal portions, which he then gives away to the entire group. Dentan explains that to express gratitude for a portion received indicates that you the kind of ungenerous person who calculates how much you give and take. "In this context saying thank you is very rude, for it suggests first that one has calculated the amount of a gift and second, that one did not expect the donor to be so generous." To call attention to one's generosity is to indicate that others are in debt to you and that you expect them to repay you. It is repugnant to egalitarian peoples even to suggest that they have been treated generously.

Richard Lee tells how he learned about this aspect of reciprocity through a revealing incident. To please the !Kung, he decided to buy a large ox and have it slaughtered as a present. After several days searching Bantu agricultural villages looking for the largest and fattest ox in the region, he acquired what appeared to be a perfect specimen. But his friends took him aside and assured him that he had been duped into buying an absolutely worthless animal. "Of course, we will eat it," they said, "but it won't fill us up—we will eat and go home to bed with stomachs rumbling." But when Lee's ox was slaughtered, it turned out to be covered with a thick layer of fat. Later, his friends explained why they had said his gift was valueless, even though they knew better than he what lay under the animal's skin:

Yes, when a young man kills much meat he comes to think of himself as a chief or big man, and he thinks of the rest of us as his servants or inferiors. We can't accept this, we refuse one who boasts, for someday his pride will make him kill somebody. So we always speak of his meat as worthless. This way we cool his heart and make him gentle.

Lee watched small groups of men and women returning home every evening with the animals and wild fruits and plants that they had killed or collected. They shared everything equally, even with campmates who had stayed behind and spent the day sleeping or taking care of their tools and weapons.

Not only do families pool that day's production, but the entire camp—residents and visitors alike—shares equally in the total quantity of food available. The evening meal of any one family is made up of portions of food from each of the other families resident. Foodstuffs are distributed raw or are prepared by the collectors and then distributed. There is a constant flow of nuts, berries, roots and melons from one family fireplace to another until each personal resident has received an equitable portion. The following morning a different combination of foragers moves out of camp and when they return late in the day, the distribution of foodstuffs is repeated.

What Hobbes did not realize is that it was in everybody's best interest in small, prestate societies to maintain each other's freedom of access to the natural habitat. Suppose a !Kung with a Hobbesian lust for power were to get up and tell his campmates, "From now on, all this land and everything on it belongs to me. I'll let you use it but only with my permission and on the condition that I get first choice of anything you capture, collect, or grow." His campmates, thinking that he had certainly gone crazy, would pack up their few belongings, take a 20- or 30-mile walk, make a new camp, and resume their usual life of egalitarian reciprocity, leaving the man who would be king alone to exercise a useless sovereignty.

To the extent that political leadership exists at all among band-and-village societies, it is exercised by individuals called headmen, who lack the power to compel others to obey their orders. But can a leader be powerless and still lead?

Spokane Museum (1952)

Ramona Wilson

These are not relics
from lost people, lost lands.
I know where they are.
Give me that digging tool.
I'll show you where
in the spring we get roots.
The wind will come, as it does,
blowing our dresses and hair
as we bend with certainty,
the pink flowers in our hands,
the earth dropping
through our fingers.

SECTION II

—

THE COMING OF
CIVILIZATION

the White Buffalo

The life of savages is so simple, and our societies are such complicated machines! The Tahitian is so close to the origin of the world, while the European is close to its old age. The contrast between them and us is greater than the difference between a newborn baby and a doddering old man. They understand absolutely nothing about our manners or our laws, and they are bound to see in them nothing but shackles disguised in a hundred different ways. Those shackles could only provoke the indignation and scorn of creatures in whom the most profound feeling is a love of liberty.

—Denis Diderot (1774)

aul Z. Simons posits civilization as virtually complete from its inception, as if domestication occurred in a kind of qualitative quantum leap. Such a provocative thesis would surely make that moment all the more compelling an object of study: the most important turning point in our history as a species.

In the clash between precivilization and the ensemble of ways to control and harness life that has all but extinguished it, what was at stake? In Shakespeare's *As You Like It*, the losing side can be seen in the exiled Duke's defense of wilderness, its "tongues in trees" and "sermons in stone and good in everything." Two-and-one-half centuries later, in a place more tangible than the Forest of Arden, Smohalla, elder of a Columbia Plateau tribe, issued a similar plaint:

"You ask me to plow the ground! Shall I take a knife and tear my mother's bosom? then when I die she will not take me to her bosom to rest.

You ask me to dig for stone! Shall I dig under her skin for her bones? then when I die I cannot enter her body to be born again. You ask me to cut grass and sell it, and be rich like white men! But how dare I cut off my mother's hair?"

The selections that follow provide some sense of the vast range of this struggle, and something of its many facets, its texture, and its fruits. Civilization's victory has had the most profound impact, both on the natural world and on our species—viscerally, culturally, and in every other way.

From Diderot in the eighteenth century, to George Marsh late in the nineteenth, up to the present in a mounting profusion and emphasis that reflects the growing crisis, we can see with increasing clarity what a truly monumental, cataclysmic watershed was the triumph of civilization. This small sampling can only suggest the scope and depth of that cataclysm.

THE EARTH AS MODIFIED BY HUMAN ACTION (1907)

GEORGE P. MARSH

Destructiveness of Man

an has too long forgotten that the earth was given to him for usufruct alone, not for consumption, still less for profligate waste. Nature has provided against the absolute destruction of any of her elementary matter, the raw material of her works; the thunderbolt and the tornado, the most convulsive throes of even the volcano and the earthquake, being only phenomena of decomposition and recomposition. But she has left it within the power of man irreparably to derange the combinations of inorganic matter and of organic life, which through the night of aeons she had been proportioning and balancing, to prepare the earth of this habitation, when in the fullness of time his Creator should call him forth to enter into its possession.

Apart from the hostile influence of man, the organic and the inorganic world are, as I have remarked, bound together by such mutual relations and adaptations as secure, if not the absolute permanence and equilibrium of both, a long continuance of the established conditions of each at any given time and place, or at least, a very slow and gradual succession of changes in those conditions. But man is everywhere a disturbing agent. Wherever he plants his foot, the harmonies of nature are turned to discords. The proportions and accommodations which insured the stability of existing arrangements are overthrown. Indigenous vegetable and animal species are extirpated, and supplanted by others of foreign origin, spontaneous production is forbidden or restricted, and the face of the earth is either laid bare or covered with a new and reluctant growth of vegetable forms and with alien tribes of animal life. These intentional changes and substitutions constitute, indeed, great revolutions; but vast as is their magnitude and importance, they are, as we shall see, insignificant in comparison with the contingent and unsought results which have flowed from them.

The fact that, of all organic beings, man alone is to be regarded as essentially a destructive power, and that he wields energies to resist which Nature—that nature whom all material life and all inorganic substance obey—is wholly impotent, tends to prove that, though living in physical nature, he is not of her, that he is of more exalted parentage, and belongs to a higher order of existences, than those which are born of her womb and live in blind submission to her dictates.

There are, indeed, brute destroyers, beasts and birds and insects of prey—all animal life feeds upon, and, of course, destroys other life—but this destruction is balanced by compensations. It is, in fact, the very means by which the existence of one tribe of animals or of vegetables is secured against being smothered by the encroachments of another; and the reproductive powers of species which serve as the food of others are always proportioned to the demand they are destined to supply. Man pursues his victims with reckless destructiveness; and while the sacrifice of life by the lower animals is limited by the cravings of appetite, he unsparingly persecutes, even to extirpation, thousands of organic forms which he can not consume.

The earth was not, in its natural condition, completely adapted to the use of man, but only to the sustenance of wild animals and wild vegetation. These live, multiply their kind in just proportion, and attain their perfect measure of strength and beauty, without producing or requiring any important change in the natural arrangements of surface or in each other's spontaneous tendencies, except such mutual repression of excessive increase as may prevent the extirpation of one species by the encroachments of another. In short, without man, lower animal and spontaneous vegetable life would have been practically constant in type, distribution and proportion, and the physical geography of the earth would have remained undisturbed for indefinite periods, and been subject to revolution only from slow development, from possible unknown cosmical causes, or from geological action.

But man, the domestic animals that serve him, the field and garden plants the products of which supply him with food and clothing, can not subsist and rise to the full development of their higher properties, unless brute and unconscious nature be effectually combated, and, in a great degree, vanquished by human art. Hence, a certain measure of transformation of terrestrial surface, of suppression of natural, and stimulation of artificially modified productivity becomes necessary. This measure man has unfortunately exceeded. He has felled the forests whose network of fibrous roots bound the mould to the rocky skeleton of the earth; but had he allowed here and there a belt of woodland to reproduce itself by spontaneous propagation, most of the mischiefs which his reckless destruction of the natural protection of the soil has occasioned would have been averted. He has broken up the mountain reservoirs, the percolation of whose waters through unseen channels supplied the fountains that refreshed his cattle and fertilized his fields; but he has neglected to maintain the cisterns and the canals of irrigation which a wise antiquity had constructed to neutralize the consequences of its own imprudence. While he has torn the thin glebe which confined the light earth of extensive plains, and has destroyed the fringe of semi-aquatic plants which skirted the coast and checked the drifting of the sea sand, he has failed to prevent the spreading of the dunes by clothing them with artificially propagated vegetation. He has ruthlessly warred on all the tribes of animated nature whose spoil he could convert to his own uses, and he has not protected the birds which prey on the insects most destructive to his own harvests.

Purely untutored humanity, it is true, interferes comparative ly little with the arrangements of nature, and the destructive agency of man becomes more and more energetic and unsparing as he advances in civilization, until the impoverishment, with which his exhaustion of the natural resources of the soil is threatening him, at last awakens him to the necessity of preserving what is left, if not of restoring what has been wantonly wasted. The wandering savage grows no cultivated vegetable, fells no forest, and extirpates no useful plant, no noxious weed. If his skill in the chase enables him to entrap numbers of the animals on which he feeds, he compensates this loss by destroying also the lion, the tiger, the wolf, the otter, the seal, and the eagle, thus indirectly protecting the feebler quadrupeds and fish and fowls, which would otherwise become the booty of beasts and birds of prey. But with stationary life, or at latest with the pastoral state, man at once commences an almost indiscriminate warfare upon all the forms of animal and vegetable existence around him, and as he advances in civilization, he gradually eradicates or transforms every spontaneous product of the soil he occupies.

Human and Brute Action Compared

It is maintained by authorities as high as any known to modern science, that the action of man upon nature, though greater in *degree*, does not differ in *kind* from that of wild animals. It is perhaps impossible to establish a radical distinction *in genere* between the two classes of effects, but there is an essential difference between the motive of action which calls out the energies of civilized man and the mere appetite which controls the life of the beast. The action of man, indeed, is frequently followed by unforeseen and undesired results, yet it is nevertheless guided by a self-conscious will aiming as often at secondary and remote as at immediate objects. The wild animal, on the other hand, acts instinctively, and, so far as we are able to perceive, always with a view to single and direct purposes. The backwoodsman and the beaver alike fell trees; the man, that he may convert the forest into an olive grove that will mature its fruit only for a succeeding generation; the beaver, that he may feed upon the bark of the trees or use them in the construction of his habitation. The action of brutes upon the material world is slow and gradual, and usually limited, in any given case, to a narrow extent of territory. Nature is allowed time and opportunity to set her restorative powers at work, and the destructive animal has hardly retired from the field of his ravages before nature has repaired the damages occasioned by his operations. In fact, he is expelled from the scene by the very efforts which she makes for the restoration of her dominion. Man, on the contrary, extends his action over vast spaces, his revolutions are swift and radical, and his devastations are, for an almost incalculable time after he has withdrawn the arm that gave the blow, irreparable.

The form of geographical surface, and very probably the climate, of a given country, depend much on the character of the vegetable life belonging to it. Man has, by domestication, greatly changed the habits and properties of the plants he rears; he has, by voluntary selection, immensely modified the forms and qualities of the animated creatures that serve him; and he has, at the same time, completely rooted out many forms of animal if not of vegetable being. What is there in the influence of brute life that corresponds to this? We have no reason to believe that, in that portion of the American continent which, though peopled by many tribes of quadruped and fowl, remained uninhabited by man or only thinly occupied by purely savage tribes, any sensible geographical change had occurred within 20 centuries before the epoch of discovery and colonization, while, during the same period, man had changed millions of square miles, in the fairest and most fertile regions of the Old World, into the barrenest deserts.

The ravages committed by man subvert the relations and destroy the balance which nature had established between her organized and her inorganic creations, and she avenges herself upon the intruder, by letting loose upon her defaced provinces destructive energies hitherto kept in check by organic forces destined to be his best auxiliaries, but which he has unwisely dispersed and driven from the field of action. When the forest is gone, the great reservoir of moisture stored up in its vegetable mould is evaporated, and returns only in deluges of rain to wash away the parched dust into which that mould has been converted. The well-wooded and humid hills are turned to ridges of dry rock, which encumber the low grounds and choke the watercourses with their debris, and—except in countries favored with an equable distribution of rain through the seasons, and a moderate and regular inclination of surface—the whole earth, unless rescued by human art from the physical degradation to which it tends, becomes an assemblage of bald mountains, of barren, turfless hills, and of swampy and malarious plains. There are parts of Asia Minor, of Northern Africa, of Greece, and even of Alpine Europe, where the operation of causes set in action by man has brought the face of the earth to a desolation almost as complete as that of the moon; and though, within that brief space of time which we call "the historical period," they are known to have been covered with luxuriant woods, verdant pastures, and fertile meadows, they are now too far deteriorated to be reclaimable by man, nor can they become again fitted for human use, except through great geological changes, or other mysterious influences or agencies of which we have no present knowledge and over which we have no prospective control. The earth is fast becoming an unfit home for its noblest inhabitant, and another era of equal human crime and human improvidence, and of like duration with that through which traces of that crime and that improvidence extend, would reduce it to such a condition of impoverished productiveness, of shattered surface, of climatic excess, as to threaten the depravation, barbarism and perhaps even extinction of the species.

Beyond Geography: The Western Spirit Against the Wilderness (1980)

Frederick Turner

꒰orth Americans are urged to think with a certain cautious fondness on our single great parable of intermingling, the marriage between the white Englishman John Rolfe and the Indian princess Pocahantas. This merging is supposed to symbolize a new and hopeful beginning in the New World. But the records of that arrangement tell us something else: the marriage was not based on any true desire for merging but rather on political expediency and a fear that no marriage could bridge. John Rolfe may, as Perry Miller says, have been horrified to find some genuine affection for Pocahantas overtaking him. God, he knew, had forbidden intermarriage to the Israelites in the interests of tribal solidarity and religious purity, and so Rolfe knew that he must find a means of justifying what otherwise could only be construed as a slide into temptation. So he did, professing and maybe believing at last, that he had entered into this marriage "for the good of this plantation, for the honour of our countrie, for the glory of God, for my owne salvation, and for converting to the true knowledge of God and Jesus Christ, an unbelieving creature." On any other terms the marriage would be a wretched instigation "hatched by him who seeketh and delighteth in man's destruction."

It was a difficult business, this possessing while withholding themselves, and the records show that it was not uniformly accomplished, though European governments would do their best by sending out shiploads of white women to keep the white men in the clearings and out of the woods. For neither the French nor the Spanish was John Rolfe a paragon, though it was the English themselves who strove most to avoid his example.

From their earliest days in the New World the English evinced an official desire both stern and shrill to keep the colonists together and to keep the natives at a safe distance. So they recorded with a sour satisfaction the fate of a splinter colony at Wessagusset, Massachusetts, that strayed off and kept Indian women but that managed itself so improvidently that it soon came near to starvation

and then became "so base" as actually to serve the Indians for sustenance. One poor, half-starved fellow was found dead where he had
stood in tidal mud in his last search for shellfish.

In Virginia, Massachusetts, and Connecticut there were stiff
penalties for those who moved ahead of the line of settlement and
lived surrounded only by woods, animals, and Indians. Such individuals, it was reasoned, would soon, like the untended olive,
turn wild themselves, and one of Cotton Mather's most persistent
laments was the "Indianizing" of the whites. As he watched them
straggle off into the woods, establish barbarous little outposts,
and settle slowly into native ways, he predicted divine retribution
as of old.

What all this tells us is that these newcomers had felt obscurely
the lure of the New World and that some had responded to it out
of old dreams long repressed in the march of Christian history. For
the fear of possession does not make sense by itself, as wilderness
travelers Joseph Conrad and Carl Jung knew so well. Only when the
fear is seen as the obverse of desire does it reveal its full truth. Conrad, witness to the European invasion of the tropics, wrote of those
who had succumbed to the temptation and had become lost forever
like Willems in *An Outcast of the Islands* and Kurtz in *The Heart of
Darkness*, who pronounced judgment on his own reversion: "The
horror! The horror!"...

Almost never do our raw wilderness documents speak openly of
this lure, of the human need to feel possessed, body and spirit, by a
landscape both visible and numinous. Only negatively can they do
so, guiding us in their obsessive loathing of those who did succumb.
And only this situation can adequately explain the strange persistence and perpetuation through the centuries of the tradition of the
willing captive: one who, being captured, refused redemption or ran
away after being redeemed, slipping off from the settlements to
return to the tribes and the wilderness. They were variously called
renegades, white Indians, squawmen, or simply degenerates, and we
might imagine that their shadowy careers would have been con-
signed to a silence beyond obloquy. But it is not so. Instead, their
stories have come down to us, loaded, true enough, with all the
weight that can be assigned scapegoats, but they have survived, and
this is significant. Never as popular because never as useful (even as
cautionary tales) as the orthodox captivity narratives, their persistence tells us of other visions of New World contact that haunted
the whites even in the very midst of their victories.

The stories begin with what was for us the beginning of the Americas, with Columbus. One Miguel Diaz of the second voyage wounded a fellow Christian in an argument and to escape punishment ran away into the woods where he became the consort of the local chieftainess. Like most of his kind, history loses track of him here, but his defection was thought worthy of record in his own time, a strange footnote to conquest.

Then, following like an apparition the traces of further conquests, there are the renegades of Cortes, Narvaez, and Soto. Though, as we have seen, Gonzalo Guerrero was not with Cortes as he pushed off from Cuba for an unknown empire, one would have thought from the diligence with which that driven leader searched for Guerrero that he was indispensable to the success of the expedition. Maybe in some way he was.

With Narvaez there were at least two who preferred to remain with the Indians rather than wander on into unknown dangers, guided only by the hope of coming out once again into Christian territory. Doroteo Teodoro, a Greek, went inland with the Indians and never came out again, though years later Soto's men would catch word out of the thicket that he still lived with his adopted people. And Lope de Oviedo ("our strongest man") turned away from the entreaties of Vaca to escape and stayed on with the natives while Vaca and the others slipped northward in the night.

On the baffled and trackless Soto expedition there were numerous deserters, notably (from a white point of view) Francisco de Guzman, bastard son of a Seville hidalgo; a hidalgo named Mancano; and another named Feryada, a Levantine. And there was an unnamed Christian described by the Portuguese knight who chronicled the expedition:

> ... the Indians came in peace, and said that the Christian who remained there would not come. The Governor wrote to him, sending ink and paper, that he might answer. The purport of the letter stated his determination to leave Florida, reminded him of his being a Christian, and that he was unwilling to leave him among heathen; that he would pardon the error he had committed in going to the Indians, should he return; and that if they should wish to detain him, to let the Governor know by writing. The Indian who took the letter came back, bringing no other response than the name and the rubric of the person written on the back, to signify that he was alive.

We have record too of several black slave defectors, notably "Carlos" and "Gomez," both of whom are known to have lived among the Indians for many years. These were presumed to have so little stake in civilization that their actions could more easily be disregarded by their Spanish masters.

Not so easily dealt with, or even entertained, was the possibility that some members of Raleigh's Lost Colony had survived by merging and migrating inland with Manteo's people. Almost two centuries after John White and his party had panted through the sandy seaside woods in much-belated relief of the colonists and had come upon that tree with its interrupted message, the old sentinel trunk was still being pointed out to the curious as evidence of that tragedy. Expeditions were compulsively sent out in 1602, 1608 and 1610 to find some trace of the colonists, even if fatal, but all failed since they could not see the clues. In 1654 friendly Indians showed another expedition evidence that some of the missing colonists had revisited the site of the old fort, but so unthinkable was the alternative of survival through merging that this hint too was left unexamined.

So, for the same reason, were other hints, before and after, including those from John Smith who had knowledge of whites living somewhere inland from Chesapeake Bay, and from a German traveler, John Ledered, who heard in North Carolina of a nation of bearded men living some miles southwest of where he was. Still the hints of survival persisted and in the eighteenth century were so strong that they compelled North Carolina's pioneer naturalist/historian John Lawson to acknowledge them. The Indians of this place tell us, he notes, "that several of their ancestors were white people and could talk in a book as we do; the truth of which is confirmed by the gray eyes being frequently found amongst these Indians and no others." But Lawson could not let the spectacle of miscegenation pass without censure. The evidence of it drove him to remark that the colony had

> miscarried for want of timely supplies from England; or through the treachery of the natives, for we may reasonably suppose that the English were forced to cohabit with them for relief and conservation; and that in the process of time they conformed themselves to the manners of their Indian relations; and thus we see how apt human nature is to degenerate.

It was only in the last decade of the nineteenth century when the menace of merging was dead that a white historian, Stephen B. Weeks, could put all this information together and make the appropriate inference. And this was that the 1587 colonists, being destroyed piecemeal by the vengeful natives of the mainland, took what they could carry and went to Manteo's people on Croatoan. Together these two groups left Croatoan, which was only a seasonal home for the natives, and moved inland by slow stages, away from the hostiles of the Outer Banks area. They were subsequently encountered on the Lumber (Lumbee) River in the mid-eighteenth century when Scots and Huguenot settlers pushed into the area. And their descendants are there to this day in what is now Robeson County. Here one may meet striking mixed bloods who trace themselves back to that intermingling on the Outer Banks and who carry on the names of the Lost Colonists.

Even while the search for the Lost Colonists was vainly going on, others of the English were becoming "lost" under the very eyes of the orthodox, and of these none excited greater opprobrium and more continuing vengeance than Thomas Morton of Massachusetts. With the single exception of a long section detailing the bestiality trial and execution of one Thomas Granger (another horrid example of the temptation to mix in the wilderness), there is no more luminous passage in Governor Bradford's long chronicle of the early years of Massachusetts than that on Morton. This is so because no one transgressed so flagrantly and almost joyously against the powerful taboo that kept the English from mingling with all that surrounded them.

The man had come out from England in 1622. By the summer of 1626 he had usurped his partner's prerogatives in a settlement venture near the present town of Quincy and was master of a heterogeneous group of indentured servants and local Indians. It was not merely that Morton was trading with the Indians and supplying them with guns, powder and spirits, for by Bradford's admission as well as by other contemporaneous accounts we know that Morton was by no means alone in this practice. Indeed, had all those English been Mortons, there might have been little to fear from selling the natives what would then have been merely hunting arms. What deeply rankled was that Morton was actively encouraging the intermingling of whites and Indians and that in doing this he was accomplishing what was so much feared: the Americanization of the English.

In the spring of 1627 Morton presided over the erection of a Maypole at the place he suggestively and salaciously styled "Mare-Mount" or "Ma-re-Mount." It was too much. A party under the command of the doughty Miles Standish broke in upon the mongrels, arrested Morton, and subsequently deported him, hoping that this would be the end of it. But the man would come back again and again to the scene of his transgression and his triumph. And again and again the English authorities would hound and persecute him. Throughout the New England winter of 1644–45 they kept the old man in a drafty jail, in irons and without charges, and when at last they let him loose he was broken. When he died "old and crazy," the English were satisfied that this threat no longer existed.

But they could not lay to rest that larger threat of which Morton had been but a particular carrier, for it kept appearing. The wilderness that had spawned it would recede by the year and mile, and with it the Indians, but all along the gnawing frontier where contact was still to be had there was the profoundly disturbing and puzzling phenomenon of "indianization." On the other hand, there were but few examples of Indians who had volunteered to go white and who had remained so. And even the missionaries, to say nothing of their lay captors, seemed ashamed of the pathetic show these converts made.

Observing a prisoner exchange between the Iroquois and the French in upper New York in 1699, Cadwallader Colden is blunt: "... notwithstanding the French Commissioners took all the Pains possible to carry Home the French, that were Prisoners with the Five Nations, and they had full Liberty from the Indians, few of them could be persuaded to return." Nor, he has to admit, is this merely a reflection on the quality of French colonial life, "for the English had as much Difficulty" in persuading their redeemed to come home, despite what Colden would claim were the obvious superiority of English ways:

No Arguments, no Intreaties, nor Tears of their Friends and Relations, could persuade many of them to leave their new Indian Friends and Acquaintance; several of them that were by the Caressings of their Relations persuaded to come Home, in a little Time grew tired of our Manner of living, and run away again to the Indians, and ended their Days with them. On the other Hand, Indian Children have been carefully educated among the English, cloathed and taught, yet, I think, there is not one

Instance, that any of these, after they had Liberty to go among their own People, and were come to Age, would remain with the English, but returned to their own Nations, and became as fond of the Indian Manner of Life as those that knew nothing of a civilized Manner of Living.

And, he concludes, what he says of this particular prisoner exchange "has been found true on many other Occasions."

Benjamin Franklin was even more pointed: When an Indian child is raised in white civilization, he remarks, the civilizing somehow does not stick, and at the first opportunity he will go back to his red relations, from whence there is no hope whatever of redeeming him. But

> when white persons of either sex have been taken prisoners young by the Indians, and have lived a while among them, tho' ransomed by their Friends, and treated with all imaginable tenderness to prevail with them to stay among the English, yet in a Short time they become disgusted with our manner of life, and the care and pains that are necessary to support it, and take the first good Opportunity of escaping again into the Woods, from whence there is no reclaiming them.

Colden's New York neighbor Crevecoeur, for all his subsequently celebrated prating about this new person who was an American, almost unwittingly reveals in the latter portion of his *Letters From an American Farmer* that the only *really* new persons are those who have forsaken white civilization for the tribes. "As long as we keep ourselves busy tilling the earth," he writes, "there is no fear of any of us becoming wild." And yet, conditions being what they then were, it was not that simple. It was not always possible to keep one's head looking down at the soil shearing away from the bright plow blade. There was always the great woods, and the life to be lived within it was, Crevecoeur admits, "singularly captivating," perhaps even superior to that so boasted of by the transplanted Europeans. For, as many knew to their rueful amazement, "thousands of Europeans are Indians, and we have no examples of even one of those aborigines having from choice become Europeans!"

The Invasion Within: The Contest of Cultures in Colonial North America (1985)

James Axtell

ven when the colonists allowed the native farmers to proceed unmolested, the English ideology of farming clashed with traditional Indian practices and beliefs. In English eyes a farm was the private property of an individual husbandman who fenced and hedged its boundaries to declare his ownership. Its products were the fruits of none but his own labor, and he could dispose of them as he saw fit. Its sole use he reserved for himself by a civil law of trespass. The Indians of the Northeast, on the other hand, held a communal idea of the land. Individual families, represented by their women, might "own" personal gardens, but fields of corn, beans and squash were the possession of the whole village which tilled them and shared equally their fruits. In the absence of corn-craving animals, the natives felt no need to fence their acres, an effort that in any event would have been wasted in their periodic removes to new ground. The southern Indians thought the colonists "childish" for confining their improvements, "as if the crop would eat itself." In the north, where denser European settlement and more numerous cattle posed a constant threat, fences often became an unwelcome necessity. Yet even there the Indians fences' were often described as "miserable" because they did not have the "heart to fence their grounds."

Before the Indians could subsist solely upon the fruits of farming they had to jettison their traditional communal ethic of sharing and hospitality. For unless everyone in the community could shift *at the same time* to an individualistic, every-man-for-himself philosophy of accumulation, the provident exemplar of the "civilized way" would lose all incentive by being eaten out of house and home by his improvident friends and relatives. Such a shift was doubly difficult to effect because the missionaries themselves preached a gospel message of charity that seemed to deny the meaner spirit of capitalism. Samuel Kirkland's encouragement of the Oneidas to a life of farming, for example, was met with precious little success and the entreaties of his impoverished villagers for food, requests that he could not in conscience deny. For as they reminded him, "You have often exhorted us to be charitable to our Neighbours and hospitable to our Foreign Brethren, when they came this way, and if it be near the end of the week invite them to tarry over the sabbath, and hear Christ's gospel, or *good news*." By sharing what they had with those in need, the Indians seemed to most Englishmen to have "no care for the future," to be squandering the wealth of the earth upon the lazy and shiftless. But in the eyes of some, "they appeared to be fulfilling the scriptures beyond those who profess to believe them, in that of taking no thought of tomorrow." It was this double obstacle—the ingrained communalism of the natives and the inconsistent philosophies of the missionaries—that in the end prevented farming from becoming the economic salvation of the Indians in the colonial period.

The English failure to reduce the natives to industry can have perhaps no more eloquent epitaph than the petition of the Mohegan Indians to the Assembly of Connecticut in 1789. Nostalgic for a golden past, their plea also bears witness to the powerful bonds of community that sustained them through almost two centuries of forced acculturation, political domination and religious intolerance, but now could sustain them no longer. "The Times are Exceedingly alter'd," they wrote.

> Yea the Times have turn'd everything Upside down, or rather we have Chang'd the good Times, Chiefly by the help of the White People. For in Times past our Fore-Fathers lived in Peace, Love and great harmony, and had everything in Great planty. When they Wanted meat they would just run into the Bush a little ways with their Weapons and would Soon bring home good venison, Racoon, Bear and Fowl. If they Choose to have Fish, they Wo'd only go to the River or along the Sea Shore and they wou'd presently fill their Cannoous With Veriety of Fish, both Scaled and shell Fish, and they had abundance of Nuts, Wild Fruit, Ground Nuts and Ground Beans, and they planted but little Corn and Beans and they kept no Cattle or Horses for they needed none. And they had no Contention about their Lands, it lay in Common to them all, and they had but one large Dish and they Cou'd all eat together in Peace and Love—But alas, it is not so now, all our Fishing, Hunting and Fowling is intirely gone. And we have now begun to Work on our Land, keep Cattle, Horses and Hogs And We Build Houses and fence in Lots, And now we

plainly See that one Dish and one Fire will not do any longer for us—Some few there are Stronger than others and they will keep off the poor, weake, the halt and the Blind, And Will take the Dish to themselves. Yea, they will rather Call White People and Molattoes to eat With them out of our Dish, and poor Widows and Orphans must be pushed one side and there they Must Set a Crying, Starving and die.

So it was with "Hearts full of Sorrow and Grief" that they asked "That our Dish of Suckuttush may be equally divided amongst us, that every one may have his own little dish by himself, that he may eat Quietly and do With his Dish as he pleases; and let every one have his own Fire." Other tribes as well eventually succumbed to the necessities of a foreign economy and an alien work ethic, but seldom before the colonies gained their independence and began to hurl a missionary onslaught at the remains of native "idleness" and independence. But even then the time clock and the regular paycheck could not completely stifle the traditional Indian values of social generosity and personal liberty.

Elements of Refusal (1988)

John Zerzan

lthough food production by its nature includes a latent readiness for political domination and although civilizing culture was from the beginning its own propaganda machine, the changeover involved a monumental struggle. Fredy Perlman's *Against Leviathan! Against His-story!* is unrivaled on this, vastly enriching Toynbee's attention to the "internal" and "external proletariats," discontents within and without civilization. Nonetheless, along the axis from digging stick farming to plow agriculture to fully differentiated irrigation systems, an almost total genocide of gatherers and hunters was necessarily effected.

The formation and storage of surpluses are part of the domesticating will to control and make static, an aspect of the tendency to symbolize. A bulwark against the flow of nature, surplus takes the forms of herd animals and granaries. Stored grain was the earliest medium of equivalence, the oldest form of capital. Only with the appearance of wealth in the shape of storable grains do the gradations of labor and social classes proceed. While there were certainly wild grains before all this (and wild wheat, by the way, is 24 percent protein compared to 12 percent for domesticated wheat) the bias of culture makes every difference. Civilization and its cities rested as much on granaries as on symbolization.

The mystery of agriculture's origin seems even more impenetrable in light of the recent reversal of long-standing notions that the previous era was one of hostility to nature and an absence of leisure. "One could no longer assume," wrote Arme, "that early man domesticated plants and animals to escape drudgery and starvation. If anything, the contrary appeared true, and the advent of farming saw the end of innocence." For a long time, the question was "why wasn't agriculture adopted much earlier in human evolution?" More recently, we know that agriculture, in Cohen's words, "is not easier than hunting and gathering and does not provide a higher quality, more palatable, or more secure food base." Thus the consensus question now is, "why was it adopted at all?"

Many theories have been advanced, none convincingly. Childe and others argue that population increase pushed human societies into more intimate contact with other species, leading to domestication and the need to produce in order to feed the additional people. But it has been shown rather conclusively that population increase did not precede agriculture but was caused by it. "I don't see any evidence anywhere in the world," concluded Flannery, "that suggest that population pressure was responsible for the beginning of agriculture." Another theory has it that major climatic changes occurred at the end of the Pleistocene, about 11,000 years ago, which upset the old hunter-gatherer life-world and led directly to the cultivation of certain surviving staples. Recent dating methods have helped demolish this approach; no such climatic shift happened that could have forced the new mode into existence. Besides, there are scores of examples of agriculture being adopted—or refused—in every type of climate. Another major hypothesis is that agriculture was introduced via a chance discovery or invention as if it had never occurred to the species before a certain moment that, for example, food grows from sprouted seeds. It seems certain that Paleolithic humanity had a virtually inexhaustible knowledge of flora and fauna for many tens of thousands of years before the cultivation of plants began, which renders this theory especially weak.

Agreement with Carl Sauer's summation that, "Agriculture did not originate from a growing or chronic shortage of food" is sufficient, in fact, to dismiss virtually all originary theories that have been advanced. A remaining idea, presented by Hahn, Isaac and others, holds that food production began at base as a religious activity. This hypothesis comes closest to plausibility.

Sheep and goats, the first animals to be domesticated, are known to have been widely used in religious ceremonies, and to have been raised in enclosed meadows for sacrificial purposes. Before they were domesticated, moreover, sheep had no wool suitable for textile purposes. The main use of the hen in southeastern Asia and the eastern Mediterranean—the earliest centers of civilization—"seems to have been," according to Darby, "sacrificial or divinatory rather than alimentary." Sauer adds that the "egg laying and meat producing qualities" of tamed fowl "are relatively late consequences of their domestication." Wild cattle were fierce and dangerous; neither the docility of oxen nor the modified meat texture of such castrates could have been foreseen. Cattle were not milked until centuries after their initial captivity, and representations indicate that their first known harnessing was to wagons in religious processions.

Plants, next to be controlled, exhibit similar backgrounds so far as is known. Consider the New World examples of squash and pumpkin, used originally as ceremonial rattles. Johannessen discussed the religious and mystical motives connected with the domestication of maize, Mexico's most important crop and center of its native Neolithic religion. Likewise Anderson investigated the selection and development of distinctive types of various cultivated plants because of their magical significance. The shamans, I should add, were well-placed in positions of power to introduce agriculture via the taming and planting involved in ritual and religion, sketchily referred to above.

Though the religious explanation of the origins of agriculture has been somewhat overlooked, it brings us, in my opinion, to the very doorstep of the real explanation of the birth of production: that non-rational, cultural force of alienation which spread, in the forms of time, language, number and art, to ultimately colonize material and psychic life in agriculture. "Religion" is too narrow a conceptualization of this infection and its growth. Domination is too weighty, too all-encompassing, to have been solely conveyed by the pathology that is religion.

But the cultural values of control and uniformity that are part of religion are certainly part of agriculture, and from the beginning. Noting that strains of corn cross-pollinate very easily, Anderson studied the very primitive agriculturists of Assam, the Naga tribe, and their variety of corn that exhibited no differences from plant to plant. True to culture, showing that it is complete from the beginning of production, the Naga kept their varieties so pure "only by a fanatical adherence to an ideal type." This exemplifies the marriage of culture and production in domestication, and its inevitable progeny, repression and work.

The scrupulous tending of strains of plants finds its parallel in the domesticating of animals, which also defies natural selection and re-establishes the controllable organic world at a debased, artificial level. Like plants, animals are mere things to be manipulated; a cow, for instance, is seen as a kind of machine for converting grass into milk. Transmuted from a state of freedom to that of helpless parasites, these animals become completely dependent on man for survival. In domestic mammals, as a rule, the size of the brain becomes relatively smaller as specimens are produced that devote more energy to growth and less to activity. Placid, infantilized, typified perhaps by the sheep, most domesticated of herd animals; the remarkable intelligence of wild sheep is completely lost in their tamed counterparts. The social relationships among domestic animals are reduced to the crudest essentials. Non-reproductive parts of the life cycle are minimized, courtship is curtailed, and the animal's very capacity to recognize its own species is impaired.

Farming also created the potential for rapid environmental destruction and the domination over nature soon began to turn the green mantle that covered the birthplaces of civilization into barren and lifeless areas. "Vast regions have changed their aspect completely," estimates Zeuner, "always to quasi-drier condition, since the beginnings of the Neolithic." Deserts now occupy most of the areas where the high civilizations once flourished, and there is much historical evidence that these early formations inevitably ruined their environments.

Throughout the Mediterranean Basin and in the adjoining Near East and Asia, agriculture turned lush and hospitable lands into depleted, dry, and rocky terrain. In *Critias*, Plato described Attica as "a skeleton wasted by disease," referring to the deforestation of Greece and contrasting it to its earlier richness. Grazing by goats and sheep, the first domesticated ruminants, was a major factor in the denuding of Greece, Lebanon, and North Africa, and the desertification of the Roman and Mesopotamian empires.

Another, more immediate impact of agriculture, brought to light increasingly in recent years, involved the physical well-being of its subjects. Lee and Devore's researches show that "the diet of gathering peoples was far better than that of cultivators, that starvation is rare, that their health status was generally superior, and that there is a lower incidence of chronic disease." Conversely, Farb summarized, "Production provides an inferior diet based on a limited number of foods, is much less reliable because of blights and the vagaries of weather, and is much more costly in terms of human labor expended."

The new field of paleopathology has reached even more emphatic conclusions, stressing, as does Angel, the "sharp decline in growth and nutrition" caused by the changeover from food gathering to food production. Earlier conclusions about life span have also been revised. Although eyewitness Spanish accounts of the sixteenth century tell of Florida Indian fathers seeing their fifth generation before passing away, it was long believed that primitive people died in their 30s and 40s. Robson, Boyden and others have dispelled the confusion of longevity with life expectancy and discovered that current hunter-gatherers, barring injury and severe infection, often outlive their civilized contemporaries. During the industrial age only fairly recently did life span lengthen for the species, and it is now widely recognized that in Paleolithic times humans were long-lived animals, once certain risks were passed. DeVries is correct in his judgment that duration of life dropped sharply upon contact with civilization.

"Tuberculosis and diarrheal disease had to await the rise of farming, measles and bubonic plague the appearance of large cities," wrote Jared Diamond. Malaria, probably the single greatest killer of humanity, and nearly all other infectious diseases are the heritage of agriculture. Nutritional and degenerative diseases in general appear with the reign of domestication and culture. Cancer, coronary thrombosis, anemia, dental caries, and mental disorders are but a few of the hallmarks of agriculture; previously women gave birth with no difficulty and little or no pain.

People were far more alive in all their senses. !Kung San, report-ed R.H. Post, have heard a single-engined plane while it was still 70 miles away, and many of them can see four moons of Jupiter with the naked eye. The summary judgment of Harris and Ross, as to "an overall decline in the quality—and probably in the length—of human life among farmers as compared with earlier hunter-gatherer groups," is understated.

One of the most persistent and universal ideas is that there was once a Golden Age of innocence before history began. Hesiod, for instance, referred to the "life-sustaining soil, which yielded its copi-ous fruits unbribed by toil." Eden was clearly the home of the hunter-gatherers and the yearning expressed by the historical images of paradise must have been that of disillusioned tillers of the soil for a lost life of freedom and relative ease.

The history of civilization shows the increasing displacement of nature from human experience, characterized in part by a narrow-ing of food choices. According to Rooney, prehistoric peoples found sustenance in over 1500 species of wild plants, whereas "All civilizations," Wenke reminds us, "have been based on the cultiva-tion of one or more of just six plant species: wheat, barley, millet, rice, maize, and potatoes."

It is a striking truth that over the centuries "the number of dif-ferent edible foods which are actually eaten," Pyke points out, "has steadily dwindled." The world's population now depends for most of its subsistence on only about 20 genera of plants while their natural strains are replaced by artificial hybrids and the genetic pool of these plants becomes far less varied.

The diversity of food tends to disappear or flatten out as the proportion of manufactured foods increases. Today the very same articles of diet are distributed worldwide so that an Inuit Eskimo and an African native may soon be eating powdered milk manufactured in Wisconsin or frozen fish sticks from a single factory in Sweden. A few big multinationals such as Unilever, the world's biggest food production company, preside over a highly integrated service system in which the object is not to nourish or even to feed, but to force an ever-increasing consumption of fabricated, processed products upon the world.

When Descartes enunciated the principle that the fullest exploitation of matter to *any* use is the whole duty of man, our separation from nature was virtually complete and the stage was set for the Industrial Revolution. 350 years later this spirit lingers in the person of Jean Vorst, Curator of France's Museum of Natural History, who pronounces that our species, "because of intellect," can no longer re-cross a certain threshold of civilization and once again become part of a natural habitat. He further states, expressing perfectly the original and persevering imperialism of agriculture, "As the earth in its primitive state is not adopted to our expansion, man must shackle it to fulfill human destiny."

The early factories literally mimicked the agricultural model, indicating again that at base all mass production is farming. The natural world is to be broken and forced to work. One thinks of the mid-American prairies where settlers had to yoke six oxen to plow in order to cut through the soil for the first time. Or a scene from the 1870s in *The Octopus* by Frank Norris, in which gang-plows were driven like "a great column of field artillery" across the San Joaquin Valley, cutting 175 furrows at once.

Today the organic, what is left of it, is fully mechanized under the aegis of a few petrochemical corporations. Their artificial fertilizers, pesticides, herbicides and near-monopoly of the world's seed stock define a total environment that integrates food production from planting to consumption.

Nature and Madness (1982)

Paul Shepard

$\mathfrak{A}$ fter nearly ten thousand years of living with apprehension about food and the binding force of its psychic disablement, it is not surprising that civilized people find it difficult to understand the absence of such worries among hunting-gathering peoples, making them seem careless and imprudent. The repressed distrust of the mother and the maternal earth can then be redirected onto those blithe savages, picturing them as unfeeling for the well-being of their families and coarsely inured to hunger and the other imagined afflictions of a brutish life. This scornful fantasy is easily enough projected upon the rest of brute creation, making it easier to believe that all animals are insentient.

It is not only an abstract Mother Earth who is the victim of this psychic deformity, but all wild things. Characteristically, farmers and townsmen do not study and speculate on wild animals or "think" them in their poetic mystery and complex behaviors. With civilization, typically fewer than 20 kinds of plants and animals in one village were tended, herded, sheltered, planted, cultivated, fertilized, harvested, cured, stored, and distributed. Sacrifice and other ceremonial activity were restricted accordingly. Even the gradual broadening of agriculture to embrace many more kinds of organisms left it far short of the rich cosmos of the hunter. Civilization increased the separation between the individual and the natural world as it did the child from the mother, amplifying an attachment that could be channeled into aggression.

The farmer and his village brethren assumed an executive task of food production, storage, and distribution that would weigh heavily on them for the same reason that all executives wear out their nerves and glands: responsibility in a situation of certain failure—if not this year, then next, or the year after that. Being held responsible for things beyond their control is especially crushing for children, for whom the world may become hopelessly chaotic. They, in their chores of goat tending or other work tasks, like the adults who managed the domesticated community, were vulnerable to weather, marauders, pests, and the demons of earth and air. Blights and witherings were inevitable, bringing not only food shortage but emotional onslaughts. The judgment is guilt, for which the penalty is scarcity.

In such a world the full belly is never enough. Like the dour Yankee farmer who sees in the clear blue sky of a Vermont spring day "a damned weather-breeder," abundance would only set the mark by which shortage would be measured.

But quantity was not the only variable. As the diversity of foods diminished—the wild alternatives becoming scarcer and more distant from villages—the danger of malnutrition increased. It is widely observed that domesticated varieties of fruits and vegetables differ from their wild ancestors in carbohydrate/protein/fat ratios as well as vitamin and mineral content. Where selection is for appearance, size, storability, or even taste, some food value may be lost. Virtually all the processes that aid storage or preservation have a similar price in decline of quality. The point is that the lack of food is not the only spur to a kind of trophic obsession, but the hungers of those who are superficially well fed might also add to this general picture of chronic preoccupation with food.

The argument can be made that anything that fixates the individual's attention on food can be associated with ontogenetic regression. I mean not only the infantile impatience to eat and the whole alimentary oral-anal romance to which he is so responsive, but the typical conservatism of older children and adolescents—the first, perhaps because of a sensitivity to strong or strange new flavors; the second, because of a psychic state in which the groping for a new selfhood is partly one of recognition of codes that identify a group. Teenagers are the weakest gourmets because they have not yet achieved a confident enough identity to free themselves to develop personal preferences. The young are wary about what they eat, probably for adaptive as well as culturally functional reasons.

The young of hunter societies are probably cautious too, and certainly such cultures had a highly developed sense of food taboos. Nonetheless, the small foraging band ate dozens of kinds of flesh (including invertebrates) and scores of kinds of roots, nuts, vegetables, and leaves. The idea that this range was born of desperation is not supported by the evidence. There were certainly seasonal opportunities and choices, but apparently to be human is to be omnivorous, to show an open, experimental attitude toward what is edible, guided by an educated taste and a wide range of options. As among all peoples, what is eaten or not eaten had cultural limits among hunter-foragers, but these did not prevent somebody in a group from eating at least some parts of almost anything.

The food-producing societies that succeeded the hunter-gatherers attempted to make virtue of defect by intensifying the cultural proscriptions on what was to be eaten in a world where, for most people, there were fewer choices than their archaic ancestors enjoyed. And how was this tightening of the belt and expanded contempt achieved? It was built into the older child and adolescent. It could be frozen at that level as part of a more general developmental check. It may have been inevitable in the shift from totemic to caste thought about animals, corresponding to the change from hunting to farming, in which wild animals ceased to be used as metaphors central to personal identity, to be less involved with analogies of assimilation and incorporation. The growth of self-identity requires coming to terms with the wild and uncontrollable within. Normally the child identifies frightening feelings and ideas with specific external objects. The sensed limitations of such objects aid his attempts to control his fears. As the natural containers for these projected feelings receded with the wilderness, a lack of substitutes may have left the child less able to cope and thus more dependent, his development impaired.

Perhaps there was no more dramatic change in the transition from hunting-gathering to farming than in the kind and number of *possessions*. Among archaic people who use no beasts of burden, true possessions are few and small. What objects are owned are divided between those privately held and those in which there is a joint interest. Among the latter, such as religious objects or the carcass of a game animal, the individual shares obligations as well as benefits, but in neither case does he accumulate or seem to feel impoverished. The wariness of gifts and the lack of accumulation found in these people are not due to nomadism, for the desire would still be evident. Nor can these characteristics be explained away as a culturally conditioned materialism, as that would beg the question.

This absence of wanting belongings seems more likely to be part of a psychological dimension of human life and its modification in civilization. "Belongings" is an interesting word, referring to membership and therefore to parts of a whole. If that whole is Me, then perhaps the acquisition of mostly man-made objects can contribute in some way to my identity—a way that may compensate for some earlier means lost when people became sedentary and their world mostly man-made landscapes. Or, if objects fail to fully suffice, we want more and more, as we crave more of a pain-killing drug. In short, what is it about the domesticated civilized world that alters the concept of self so that it is enhanced by property?

My self is to some extent made by me, at least insofar as I seem to gain control over it. A wilderness environment is, on the contrary, mostly given. For the hunter-forager, this Me in a non-Me world is the most penetrating and powerful realization in life. The mature person in such a culture is not concerned with blunting that dreadful reality but with establishing lines of connectedness or relationship. Formal culture is shaped by the elaboration of covenants and negotiations with the Other. The separation makes impossible a fuzzy confusion; there is no vague "identity with nature," but rather a lifelong task of formulating—and internalizing—treaties of affiliation. The forms and terms of that relationship become part of a secondary level of my identity, the background or gestalt. This refining of what-I-am-not is a developmental matter, and the human life cycle conforms to stages in its progress.

Now consider the process in a world in which that Other has mostly disappeared. Food, tools, animals, structures, whole landscapes are man-made; even to me personally they seem more made than given and serve as extensions of that part of the self which I determine. My infantile ego glories in this great consuming I-am. Everything in sight belongs to me in the same sense as my members: legs, arms, hands, and so on. The buildings, streets, and cultivated fields are all continuous with my voluntary nervous system, my tamed, controlled self.

In the ideology of farming, wild things are enemies of the tame; the wild Other is not the context but the opponent of "my" domain. Impulses, fears, and dreams—the realm of the unconscious—no longer are represented by the community of wild things with which I can work out a meaningful relationship. The unconscious is driven deeper and away with the wilderness. New definitions of the self by trade and political subordination in part replace the metaphoric reciprocity between natural and cultural in the totemic life of the hunter-foragers. But the new system defines by exclusion. What had been a complementary entity embracing friendly and dangerous parts in a unified cosmos now takes on the colors of hostility and fragmentation. Even where the great earth religions of high agriculture tend to mend this rupture in the mythology of the symbolic mother, its stunting of the identity process remains.

Although he formulated the cognitive distinctions between totemic culture, with its analogy of a system of differences in non-human nature as a paradigm for the organization of culture, and caste or agriculture, which find models for human relationships in the types of things made, Lévi-Strauss avoided the psychological developmental implications with admirable caution. But it is clear from the developmental scheme of Erikson that fine mastery of the neuromuscular system, self-discipline of the body, the emergence of skills, and awakening to tools are late-juvenile and early-adolescent concerns. In farming, the land itself becomes a tool, an instrument of production, a possession that is at once the object and implement of vocation as well as a definer of the self.

As farming shifts from subsistence to monoculture, village specialists who do not themselves cultivate the soil appear. Their roles are psychologically and mythically reintegrated into society as a whole. Smith, potter, clerk, and priest become constituents of the new reality. That reality is for them all like the pot to the potter:

(1) the wild world has reduced significance in his own conscious identity and may therefore be perceived (along with some part of himself) as chaotic; (2) he himself, like his pot, is a static *made* object, and, by inference, so is the rest of society and the world; (3) there is a central core of nonlivingness in himself; (4) the ultimate refinements in his unique self are to be achieved by acts of will or creativity; (5) daily labor—routine, repetitive motions for long hours at a time—is at the heart of his being; (6) his relationship to others is based on an exchange of possessions, and the accumulation of them is a measure of his personal achievement; and (7) the nonhuman world is primarily a source of substance to be shaped or made by man, as it was mythically by God.

These are but fragments of the world of the artisan. Gradations exist between that world and totemic cultures. The transition took many centuries before man's concept of the wilderness was indeed defined by the first synonym in Roget's Thesaurus: "disorder." In the earliest farming societies perhaps there were only nuances of the psychology of domestication. The individual would not see himself as a possession or conceive of being possessed by others until tribal villages coalesced into chiefdoms and he was conscripted or enslaved or his labor sold as a commodity, events that may have been as much an outcome as a cause of the new consciousness. That was many generations in the future as the first harvesters of wild wheat began to save some grains to plant. Yet we see them headed, however tentatively, toward the view of the planet as a thing rather than a thou, a product instead of an organism, to be possessed rather than encountered as a presence.

This attitude connects with the psychological position of early infancy, when differentiation between the living and the nonliving is still unclear. The badly nurtured infant may become imprinted with the hardness of its cradle or bottle so irreversibly that it cannot, even as an adult, form fully caring human relationships. But that is the extreme case. The earliest farmers were inclined to represent the landscape as a living being, even, at first, to conceive life in made things. But as those things became commodities and infancy was reshaped accordingly, the cosmos would become increasingly ambiguous. Attempts to resolve this conflict between the "itness" and the numen of things—both the landscape and its reciprocal, the inner self—are a major goal of the religious and cultural activity of civilization.

The *domestication* of animals had effects on human perception that went far beyond its economic implications. Men had been observing animals closely as a major intellectual activity for several million years. They have not been deterred, even by so momentous a change in the condition of man/animal relationship as domestication, but the message has been altered. Changes in the animals themselves, brought about by captivity and breeding programs, are widely recognized. These changes include plumper and more rounded features, greater docility and submissiveness, reduced mobility, simplification of complex behaviors (such as courtship), the broadening or generalizing of signals to which social responses are given (such as following behavior), reduced hardiness, and less specialized environmental and nutritional requirements. The sum effect of these is infantilization. The new message is an emotional appeal, sense of mastery, and relative simplicity of animal life. The style conveyed as a metaphor by the wild animal is altered to literal model and metonymic subordinate: life is inevitable physical deformity and limitation, mindless frolic and alarms, bluntness, following and being herded, being fertile when called upon, representing nature at a new, cruder level.

One or another of the domesticated forms was widely used as a substitute in human relations; as slave, sexual partner, companion, caretaker, family member. Animal and human discriminations that sustained barriers between species were breached, suggesting nothing so much in human experience as the very small child's inability to see the difference between dogs and cows. Pet-keeping, virtually a civilized institution, is an abyss of covert and unconscious uses of animals in the service of psychological needs, glossed over as play and companionship. The more extremely perverted private abuse of animals grades off into the sadistic slaughter of animals in public spectacles, of which the modern bullfight is an extravagant example.

Before civilization, animals were seen as belonging to their own nation and to be the bearers of messages and gifts of meat from a sacred domain. In the village they became possessions. Yet ancient avatars, they remained fascinating in human eyes.

A select and altered little group of animals, filtered through the bottleneck of domestication, came in human experience to represent the whole of animals of value to people. The ancient human approach to consciousness by seeing—or discovering—the self through other eyes and the need to encounter the otherness of the cosmos in its kindred aspect were two of the burdens thrust upon these deformed creatures. To educate his powers of discrimination and wonder, the child, born to expect subtle and infinite possibilities, was presented with fat hulks, vicious manics, and hypertrophied drudges. The psychological introjection of these as part of the self put the child on a detour in the developmental process that would culminate in a dead end, posted "You can't get there from here."

HEALTH AND THE RISE
OF CIVILIZATION (1989)

MARK NATHAN COHEN

The earliest visible populations of prehistory nonetheless do surprisingly well if we compare them to the actual record of human history rather than to our romantic images of civilized progress. Civilization has not been as successful in guaranteeing human well-being as we like to believe, at least for most of our history. Apparently, improvements in technology and organization have not entirely offset the demands of increasing population; too many of the patterns and activities of civilized lifestyles have generated costs as well as benefits.

There is no evidence either from ethnographic accounts or archaeological excavations to suggest that rates of accidental trauma or interpersonal violence declined substantially with the adoption of more civilized forms of political organization. In fact, some evidence from archaeological sites and from historical sources suggests the opposite.

Evidence from both ethnographic descriptions of contemporary hunters and the archaeological record suggests that the major trend in the quality and quantity of human diets has been downward. Contemporary hunter-gatherers, although lean and occasionally hungry, enjoy levels of caloric intake that compare favorably with national averages for many major countries of the Third World and that are generally above those of the poor in the modern world. Even the poorest recorded hunter-gatherer group enjoys a caloric intake superior to that of impoverished contemporary urban populations. Prehistoric hunter-gatherers appear to have enjoyed richer environments and to have been better nourished than most subsequent populations (primitive and civilized alike). Whenever we can glimpse the remains of anatomically modern human beings who lived in early prehistoric environments still rich in large game, they are often relatively large people displaying comparatively few signs of qualitative malnutrition. The subsequent trend in human size and stature is irregular but is more often downward than upward in most parts of the world until the nineteenth or twentieth century.

The diets of hunter-gatherers appear to be comparatively well balanced, even when they are lean. Ethnographic accounts of contemporary groups suggest that protein intakes are commonly quite high, comparable to those of affluent modern groups and substantially above world averages. Protein deficiency is almost unknown in these groups, and vitamin and mineral deficiencies are rare and usually mild in comparison to rates reported from many Third World populations. Archaeological evidence suggests that specific deficiencies, including that of iron (anemia), vitamin D (rickets), and, more controversially, vitamin C (scurvy)—as well as such general signs of protein calorie malnutrition as childhood growth retardation—have generally become more common in history rather than declining....

Among farmers, increasing population required more and more frequent cropping of land and the use of more and more marginal soils, both of which further diminished returns for labor. This trend may or may not have been offset by such technological improvements in farming as the use of metal tools, specialization of labor, and efficiencies associated with large-scale production that tend to increase individual productivity as well as total production.

But whether the efficiency of farming increased or declined, the nutrition of individuals appears often to have declined for any of several reasons: because increasingly complex society placed new barriers between individuals and flexible access to resources, because trade often siphoned resources away, because some segments of the society increasingly had only indirect access to food, because investments in new technology to improve production focused power in the hands of elites so that their benefits were not widely shared, and perhaps because of the outright exploitation and deprivation of some segments of society. In addition, more complex societies have had to devote an increasing amount of their productive energy to intergroup competition, the maintenance of intragroup order, the celebration of the community itself, and the privilege of the elite, rather than focusing on the biological maintenance of individuals.

In any case, the popular impression that nutrition has improved through history reflects twentieth-century affluence and seems to have as much to do with class privilege as with an overall increase in productivity. Neither the lower classes of prehistoric and classical empires nor the contemporary Third World have shared in the improvement in caloric intake; consumption of animal protein seems to have declined for all but privileged groups.

There is no clear evidence that the evolution of civilization has reduced the risk of resource failure and starvation as successfully as we like to believe. Episodes of starvation occur among hunter-gatherer bands because natural resources fail and because they have limited ability either to store or to transport food. The risk of starvation is offset, in part, by the relative freedom of hunter-gatherers to move around and find new resources, but it is clear that with limited technology of transport they can move neither far nor fast enough to escape severe fluctuations in natural resources. But each of the strategies that sedentary and civilized populations use to reduce or eliminate food crises generate costs and risks as well as benefits. The supplementation of foraging economies by small-scale cultivation may help to reduce the risk of seasonal hunger, particularly in crowded and depleted environments. The manipulation and protection of species involved in farming may help to reduce the risk of crop failure. The storage of food in sedentary communities may also help protect the population against seasonal shortages or crop failure. But these advantages may be outweighed by the greater vulnerability that domestic crop species often display toward climatic fluctuations or other natural hazards, a vulnerability that is then exacerbated by the specialized nature or narrow focus of many agricultural systems. The advantages are also offset by the loss of mobility that results from agriculture and storage, the limits and failures of primitive storage systems, and the vulnerability of sedentary communities to political expropriation of their stored resources.

Although the intensification of agriculture expanded production, it may have increased risk in both natural and cultural terms by increasing the risk of soil exhaustion in central growing areas and of crop failure in marginal areas. Such investments as irrigation to maintain or increase productivity may have helped to protect the food supply, but they generated new risks of their own and introduced new kinds of instability by making production more vulnerable to economic and political forces that could disrupt or distort the pattern of investment. Similarly, specialization of production increased the range of products that could be made and increased the overall efficiency of production, but it also placed large segments of the population at the mercy of fickle systems of exchange or equally fickle social and political entitlements.

Modern storage and transport may reduce vulnerability to natural crises, but they increase vulnerability to disruption of the technological—or political and economic—basis of the storage and transport systems themselves. Transport and storage systems are difficult and expensive to maintain. Governments that have the power to move large amounts of food long distances to offset famine and the power to stimulate investment in protective systems of storage and transport also have and can exercise the power to withhold aid and divert investment. The same market mechanisms that facilitate the rapid movement of produce on a large scale, potentially helping to prevent starvation, also set up patterns of international competition in production and consumption that may threaten starvation to those individuals who depend on world markets to provide their food, an ever-increasing proportion of the world population.

It is therefore not clear, in theory, that civilization improves the reliability of the individual diet. As the data summarized in earlier chapters suggest, neither the record of ethnography and history nor that of archaeology provide any clear indication of progressive increase in the reliability (as opposed to the total size) of human food supplies with the evolution of civilization.

Similar points can be made with reference to the natural history of infectious disease. The data reviewed in preceding chapters suggest that prehistoric hunting and gathering populations would have been visited by fewer infections and suffered lower overall rates of parasitization than most other world populations, except for those of the last century, during which antibiotics have begun to offer serious protection against infection.

The major infectious diseases experienced by isolated hunting and gathering bands are likely to have been of two types: zoonotic diseases, caused by organisms whose life cycles were largely independent of human habits; and chronic diseases, handed directly from person to person, the transmission of which were unlikely to have been discouraged by small group size. Of the two categories, the zoonotic infections are undoubtedly the more important. They are likely to have been severe or even rapidly fatal because they were poorly adapted to human hosts. Moreover, zoonotic diseases may have had a substantial impact on small populations by eliminating productive adults. But in another respect their impact would have been limited because they did not pass from person to person.

By virtue of mobility and the handling of animal carcasses, hunter-gatherers are likely to have been exposed to a wider range of zoonotic infections than are more civilized populations. Mobility may also have exposed hunter-gatherers to the traveler's

diarrhea phenomenon in which local microvariants of any parasite (including zoonoses) placed repeated stress on the body's
immune response.

The chronic diseases, which can spread among small isolated
groups, appear to have been relatively unimportant, although they
undoubtedly pose a burden of disease that can often be rapidly eliminated by twentieth-century medicine. First, such chronic diseases
appear to provoke relatively little morbidity in those chronically
exposed. Moreover, the skeletal evidence suggests that even yaws
and other common low-grade infections (periostitis) associated with
infections by organisms now common to the human environment
were usually less frequent and less severe among small, early mobile
populations than among more sedentary and dense human groups.
Similar arguments appear to apply to tuberculosis and leprosy, judg-
ing from the record of the skeletons. Even though epidemiologists
now concede that tuberculosis could have spread and persisted in
small groups, the evidence suggests overwhelmingly that it is primarily a disease of dense urban populations.

Similarly, chronic intestinal infestation by bacterial, protozoan,
and helminth parasites, although displaying significant variation in
occurrence according to the natural; environment, generally appears
to be minimized by small group size and mobility. At least, the
prevalence of specific parasites and the parasite load, or size of the
individual dose, is minimized, although in some environments
mobility actually appears to have increased the variety of parasites
encountered. Ethnographic observations suggest that parasite loads
are often relatively low in mobile bands and commonly increase as
sedentary lifestyles are adopted. Similar observations imply that
intestinal infestations are commonly more severe in sedentary populations than in their more mobile neighbors. The data also indicate
that primitive populations often display better accommodation to
their indigenous parasites (that is, fewer symptoms of disease in proportion to their parasite load) than we might otherwise expect. The
archaeological evidence suggests that, insofar as intestinal parasite
loads can be measured by their effects on overall nutrition (for
example, on rates of anemia), these infections were relatively mild in
early human populations but became increasingly severe as populations grew larger and more sedentary. In one case where
comparative analysis of archaeological mummies from different
periods has been undertaken, there is direct evidence of an increase
in pathological intestinal bacteria with the adoption of sedentism.

In another case, analysis of feces has documented an increase in intestinal parasites with sedentism.

Many major vector-borne infections may also have been less important among prehistoric hunter-gatherers than they are in the modern world. The habits of vectors of such major diseases as malaria, schistosomiasis, and bubonic plague suggest that among relatively small human groups without transportation other than walking these diseases are unlikely to have provided anything like the burden of morbidity and mortality that they inflicted on historic and contemporary populations.

Epidemiological theory further predicts the failure of most epidemic diseases ever to spread in small isolated populations or in groups of moderate size connected only by transportation on foot. Moreover, studies on the blood sera of contemporary isolated groups suggest that, although small size and isolation is not a complete guarantee against the transmission of such diseases in the vicinity, the spread from group to group is at best haphazard and irregular. The pattern suggests that contemporary isolates are at risk to epidemics once the diseases are maintained by civilized populations, but it seems to confirm predictions that such diseases would and could not have flourished and spread—because they would not reliably have been transmitted—in a world inhabited entirely by small and isolated groups in which there were no civilized reservoirs of diseases and all transportation of diseases could occur only at the speed of walking human beings.

In addition, overwhelming historical evidence suggests that the greatest rates of morbidity and death from infection are associated with the introduction of new diseases from one region of the world to another by processes associated with civilized transport of goods at speeds and over distances outside the range of movements common to hunting and gathering groups. Small-scale societies move people among groups and enjoy periodic aggregation and dispersal, but they do not move the distances associated with historic and modern religious pilgrimages or military campaigns, nor do they move at the speed associated with rapid modern forms of transportation. The increase in the transportation of people and exogenous diseases seems likely to have had far more profound effects on health than the small burden of traveler's diarrhea imposed by the small-scale movements of hunter-gatherers.

Prehistoric hunting and gathering populations may also have had one other important advantage over many more civilized groups. Given the widely recognized (and generally positive or synergistic) association of malnutrition and disease, the relatively good nutrition of hunter-gatherers may further have buffered them against the infections they did encounter.

In any case, the record of the skeletons appears to suggest that severe episodes of stress that disrupted the growth of children (acute episodes of infection or epidemics and/or episodes of resource failure and starvation) did not decline—and if anything became increasingly common with the evolution of civilization in prehistory....

There is also evidence, primarily from ethnographic sources, that primitive populations suffer relatively low rates of many degenerative diseases compared, at least, to the more affluent of modern societies, even after corrections are made for the different distribution of adult ages. Primitive populations (hunter-gatherers, subsistence farmers, and all groups who do not subsist on modern refined foods) appear to enjoy several nutritional advantages over more affluent modern societies that protect them from many of the diseases that now afflict us. High bulk diets, diets with relatively few calories in proportion to other nutrients, diets low in total fat (and particularly low in saturated fat), and diets high in potassium and low in sodium, which are common to such groups, appear to help protect them against a series of degenerative conditions that plague the more affluent of modern populations, often in proportion to their affluence. Diabetes mellitus appears to be extremely rare in primitive groups (both hunter-gatherers and farmers) as are circulatory problems, including high blood pressure, heart disease, and strokes. Similarly, disorders associated with poor bowel function, such as appendicitis, diverticulosis, hiatal hernia, varicose veins, hemorrhoids, and bowel cancers, appear rare. Rates of many other types of cancer—particularly breast and lung—appear to be low in most small-scale societies, even when corrected for the small proportion of elderly often observed; even those cancers that we now consider to be diseases of under-development, such as Burkitt's lymphoma and cancer of the liver, may be the historical product of changes in human behavior involving food storage or the human-assisted spread of vector-borne infections. The record of the skeletons suggests, through the scarcity of metastases in bone, that cancers were comparatively rare in prehistory.

The history of human life expectancy is much harder to describe or summarize with any precision because the evidence is so fragmentary and so many controversies are involved in its interpretation.

But once we look beyond the very high life expectancies of mid-twentieth century affluent nations, the existing data also appear to suggest a pattern that is both more complex and less progressive than we are accustomed to believe.

Contrary to assumptions once widely held, the slow growth of prehistoric populations need not imply exceedingly high rates of mortality. Evidence of low fertility and/or the use of birth control by small-scale groups suggests (if we use modern life tables) that average rates of population growth very near zero could have been maintained by groups suffering only historically moderate mortality (life expectancy of 25 to 30 years at birth with 50 to 60 percent of infants reaching adulthood—figures that appear to match those observed in ethnographic and archaeological samples) that would have balanced fertility, which was probably below the averages of more sedentary modern populations. The prehistoric acceleration of population growth after the adoption of sedentism and farming, if it is not an artifact of archaeological reconstruction, could be explained by an increase in fertility or altered birth control decisions that appear to accompany sedentism and agriculture. This explanation fits the available data better than any competing hypothesis.

It is not clear whether the adoption of sedentism or farming would have increased or decreased the proportion of individuals dying as infants or children. The advantages of sedentism may have been offset by risks associated with increased infection, closer spacing of children, or the substitution of starchy gruels for mother's milk and other more nutritious weaning foods. The intensification of agriculture and the adoption of more civilized lifestyles may not have improved the probability of surviving childhood until quite recently. Rates of infant and child mortality observed in the smallest contemporary groups (or reconstructed with less certainty among prehistoric groups) would not have embarrassed most European countries until sometime in the nineteenth century and were, in fact, superior to urban rates of child mortality through most of the nineteenth century (and much of the twentieth century in many Third World cities).

There is no evidence from archaeological samples to suggest that adult life expectancy increased with the adoption of sedentism or farming; there is some evidence (complicated by the effects of a probably acceleration of population growth on cemetery samples) to suggest that adult life expectancy may actually have declined as farming was adopted. In later stages of the intensification of agriculture and the development of civilization, adult life expectancy most often increased—and often increased substantially—but the trend was spottier than we sometimes realize. Archaeological populations from the Iron Age or even the Medieval period in Europe and the Middle East or from the Mississippian period in North America often suggest average adult ages at death in the middle or upper thirties, not substantially different from (and sometimes lower than) those of the earliest visible populations in the same regions. Moreover, the historic improvement in adult life expectancy may have resulted at least in part from increasing infant and child mortality and the consequent "select" nature of those entering adulthood as epidemic diseases shifted their focus from adults to children....

These data clearly imply that we need to rethink both scholarly and popular images of human progress and cultural evolution. We have built our images of human history too exclusively from the experiences of privileged classes and populations, and we have assumed too close a fit between technological advances and progress for individual lives.

In scholarly terms, these data—which often suggest diminishing returns to health and nutrition—tend to undermine models of cultural evolution based on technological advances. They add weight to theories of cultural evolution that emphasize environmental constraints, demographic pressure, and competition and social exploitation, rather than technological or social progress, as the primary instigators of social change.... Similarly, the archaeological evidence that outlying populations often suffered reduced health as a consequence of their inclusion in larger political units, the clear class stratification of health in early and modern civilizations, and the general failure of either early or modern civilizations to promote clear improvements in health, nutrition, or economic homeostasis for large segments of their populations until the very recent past all reinforce competitive and exploitative models of the origins and function of civilized states.

In popular terms, I think that we must substantially revise our traditional sense that civilization represents progress in human well-being—or at least that it did so for most people for most of history prior to the twentieth century. The comparative data simply do not support that image.

THE SEARCH FOR SOCIETY (1989)

ROBIN FOX

Since the beginnings of civilization we have known that something was wrong: since the Book of the Dead, since the Mahabharata, since Sophocles and Aeschylus, since the Book of Ecclesiastes. It has been variously diagnosed: the lust for knowledge of the Judaic first parents; the hubris of the Greeks; the Christian sin of pride; the Confucian disharmony with nature; the Hindu/Buddhist overvaluation of existence. Various remedies have been proposed: the Judaic obedience; the Greek stoicism; the Christian brotherhood of man in Christ; the Confucian cultivation of harmony; the Buddhist recognition of the oneness of existence, and eventual freedom from its determinacy. None of them has worked. (Or as the cynic would have it, none of them has been tried.)

The nineteenth century advanced the doctrine of inevitable progress allied to its eighteenth-century legacy of faith in reason and human perfectibility through education. We thought, for a brief period ("recent history"!) that we could do anything. We can't. But it comes hard to our egos to accept limitations after centuries of "progress." Will we learn to read those centuries as mere blips on the evolutionary trajectory? As aberrantly wild swings of the pendulum? As going too far? Will we come to understand that consciousness can only exist out of context for so long before it rebels against its unnatural exile? We might, given some terrible shock to the body social of the species, as Marx envisioned in his way. (Thus returning us to our state of *Gattungswesen*—species-being—where we existed before the Greek invention of the polis cut us off from nature in the first great act of alienation.)

My Name is Chellis and I'm in Recovery from Western Civilization (1994)

Chellis Glendinning

The emergence of this infirmity had been a long time coming, in slow and continual evolution ever since the initiation of a psychic and ecological development some ten thousand years before. This historic development, the launching of the neolithic, was an occurrence that began penetrating the human mind the moment we purposefully isolated domestic plants from natural ones, the moment we captured beasts from their homes in the wild and corralled them into human-built enclosures. Previous to this event humans had indeed participated in the evolution of the natural world—carrying seeds, through the wilderness, dropping, scattering, or planting them, returning later to harvest them; hunting animals by building branch and rock obstructions; catching fish and insects; constructing temporary shelters out of rock, trees, and ice. But this development was something different, something unprecedented. This was the purposeful separation of human existence from the rest of life: the domestication of the human species. To Paul Shepard's mind, the original dualism—the tame/wild dichotomy—came into being, and with it, the elliptical wholeness of the world was clipped.

The fence was the ultimate symbol of this development. What came to reside within its confines—domesticated cereals, cultivated flowers, oxen, permanent housing structures—was said to be tame; to be valued, controlled, and identified with. What existed outside was wild—"weeds," weather, wind, the woods—perennially threatening human survival; to be feared, scorned, and kept at bay. This dichotomy has since crystallized and come to define our lives with the myriads of fences separating us from the wild world and the myriads of fencelike artifacts and practices we have come to accept as "the way things are": economic individualism, private property, exclusive rights, nation-states, resource wars, nuclear missiles—until today our civilization has nearly succeeded at domesticating the entire planet and is looking, in the near future, to enclose both the outer space of other planets and the inner space of our own minds, genes, and molecules.

"Separation," writes feminist philosopher Susan Griffin of this phenomenon. "The clean from the unclean. The decaying, the putrid, the polluted, the fetid, the eroded, waste, defecation, from the unchanging.... The errant from the city. The ghetto. The ghetto of Jews. The ghetto of Moors. The quarter of prostitutes. The ghetto of blacks. The neighborhood of lesbians. The prison. The witch house. The underworld. The underground. The sewer. Space divided. The inch. The foot. The mile. The boundary. The border. The nation. The promised land. The chosen ones. The prophets, the elect, the vanguard, the sanctified, the canonized, and the canonizers."

In the psychotherapeutic process, one assumption mental-health professionals consistently make is that whatever behavior, feeling, or state of consciousness a person experiences, expresses, or presents exists for a reason. A good reason. If you and I were given the task of acting as psychotherapists for this domesticated world, we would immediately focus our attention on the "presenting symptom" of separation and divisiveness. We might wonder if the overwhelming success of linear perspective as the sole definition of visual reality isn't a symptom of some deeper condition seeking expression. And we might ask: why did some humans create—and then rationalize with elaborate devices, ideologies, and defenses—an unprecedented way of seeing the world that is based on distancing and detachment?

For a clue, we might look to survivors of post-traumatic stress disorder: Vietnam veterans, rape victims and survivors of childhood abuse, sufferers of both natural and technology-induced disasters. One of the most common symptoms to manifest itself after the experience of trauma is the neurophysiological response of disembodiment—"leaving one's body" to escape from pain that is literally too overwhelming to bear. Some people who have endured traumatic events, in describing the experience, tell of a sensation of "lifting out of their bodies," of watching the event from a vantage point slightly above, a vantage point not unlike that of linear perspective. Others tell of escaping into a post-trauma state of mental activity devoid of feeling or body awareness, a state not unlike that considered "normal" in today's dominant culture and taught in our schools and universities.

As psychotherapists, we might eventually wonder and ask: could it be that our very culture splits mind from body, intellect from feeling, because we as individuals are suffering from post-traumatic stress?

Could it be that we as individuals are dissociated because we inhabit a culture that is founded on and perpetrates traumatic stress?

Could it be that the linear perspective that infuses our vision—from our glorification of intellectual distancing to our debunking of the earthier realms of feeling and intuition; to our relentless "lifting" upward with skyscrapers and space shuttles; to the ultimate techno-utopian vision of "downloading" human knowledge into self-perpetuating computers to make embodied life obsolete—that such a perception is the result of some traumatic violation that happened in our human past?

Mythologies describing pre-agricultural times from cultures as divergent as African, Native American, and Hebraic tell of human beings at one time living in balance on the Earth. The western world claims at least five traditions that describe an earlier, better period: the Hebrew Garden of Eden, the Sumerian Dilum, the Iranian Garden of Yima, the Egyptian Tep Zepi, and the Greek Golden Age. Ovid's words in *Metamorphoses* are among the most cited and most revealing.

> Penalties and fears there were none, nor were threatening words inscribed on unchanging bronze; nor did the suppliant crowd fear the words of its judge, but they were safe without protectors. Not yet did the pine cut from its mountain tops descend into the flowing waters to visit foreign lands, nor did deep trenches gird the town, nor were there straight trumpets, nor horns of twisted brass, nor helmets, nor swords. Without the use of soldiers the peoples in safety enjoyed their sweet repose. Earth herself, unburdened and untouched by the hoe and unwounded by the ploughshare, gave all things freely.

Most of these mythic legends go on to tell of a "fall" consistently depicted as a lowering of the quality of human character and culture. In recent decades such stories may have appeared to us as quaint allegories, bedtime stories, or the stuff of a good film. But today, from our situs within the psychological and ecological crises of western civilization, these stories become dreams so transparent we barely need to interpret them. According to myths of the Bantu of southern Africa, God was driven away from the Earth *by humanity's insensitivity to nature*. The Yurok of northern California say that at a certain point in history, *people disrupted nature's balance* with their greed. The Biblical story of Eden tells of a great Fall when Adam and Eve *removed themselves from "the Garden"* and came to know evil.

In his work with survivors of post-traumatic stress, psychotherapist and author Terry Kellogg emphasizes the fact that abusive behaviors—whether we direct them toward ourselves, other people, or other species—are not natural to human beings. People enact such behaviors because *something unnatural has happened to them* and they have become damaged. With this important insight in mind, we might consider that the "fall" described in myths around the world was not a preordained event destined to occur in the unfoldment of human consciousness, as some linear-progressive New Age thinkers posit; nor was it the result of what the Bible terms "original sin," which carries with it the onus of fault and blame. We might consider that this historic alteration in our nature, or at least in how we express our nature, came about as the result of *something unnatural that happened to us*.

What could this "something" be?

Because we are creatures who were born to live in vital participation with the natural world, the violation of this participation forms the basis of our *original trauma*. This is the systemic removal of our lives from our previously assumed elliptical participation in nature's world—from the tendrils of earthy textures, the seasons of sun and stars, carrying our babies across rivers, hunting the sacred game, the power of the life force. It is a severance that in the western world was initiated slowly and subtly at first with the domestication of plants and animals, grew in intensity with the emergence of large-scale civilizations, and has developed to pathological proportion with mass technological society—until today you and I can actually live for a week or a month without smelling a tree, witnessing the passage of the moon, or meeting an animal in the wild, much less knowing the spirits of these beings or fathoming the interconnections between their destinies and our own. Original trauma is the disorientation we experience, however consciously or unconsciously, because we do not live in the natural world. It is the psychic displacement, the exile, that is inherent in civilized life. It is our homelessness.

SOCIETY AGAINST THE STATE (1977)

PIERRE CLASTRES

In conclusion, let us return to the exemplary world of the Tupi-Guarani. Here is a society that was encroached upon, threatened, by the irresistible rise of the chiefs; it responded by calling up from within itself and releasing forces capable, albeit at the price of collective near suicide, of thwarting the dynamic of the chieftainship, of cutting short the movement that might have caused it to transform the chiefs into law-giving kings. On one side, the chiefs, on the other, and standing against them, the prophets: these were the essential lines of Tupi-Guarani society at the end of the fifteenth century. And the prophetic "machine" worked perfectly well, since the *karai* were able to sweep astonishing masses of Indians along behind them, so spellbound (as one would say today) by the language of those men that they would accompany them to the point of death.

What is the significance of all that? Armed only with their Word, the prophets were able to bring about a "mobilization" of the Indians; they were able to accomplish that impossible thing in primitive society: to unify, in the religious migration, the multifarious variety of the tribes. They managed to carry out the whole "program" of the chiefs with a single stroke. Was this the ruse of history? A fatal flaw that, in spite of everything, dooms primitive society to dependency? There is no way of knowing. But, in any case, the insurrectional act of the prophets against the chiefs conferred on the former, through a strange reversal of things, infinitely more power than was held by the latter. So perhaps the idea of the spoken word being opposed to violence needs to be amended. While the primitive chief is under the obligation of *innocent* speech, primitive society can also, given quite specific conditions, lend its ear to another sort of speech, forgetting that it is uttered like a commandment: prophecy is that other speech. In the discourse of the prophets there may lie the seeds of the discourse of power, and beneath the exalted features of the mover of men, the one who tells them of their desire, the silent figure of the Despot may be hiding.

Prophetic speech, the power of that speech: might this be the place where power *tout court* originated, the beginning of the State in the Word? Prophets who were soul-winners before they were the masters of men? Perhaps. But even in the extreme experience of prophetism (extreme in that the Tupi-Guarani society had doubtless reached, whether for demographic reasons or others, the furthest limits that define a society as primitive), what the Savages exhibit is the continual effort to prevent chiefs from being chiefs, the refusal of unification, the endeavor to exorcise the One, the State. It is said that the history of peoples who have a history is the history of class struggle. It might be said, with at least as much truthfulness, that the history of peoples without history is the history of their struggle against the State.

The Land of the Naked People (2003)

Madhusree Mukerjee

2l plant that the Onge used to reduce fever has been found to kill the exceedingly dangerous cerebral malarial virus. The plant's name remains secret, especially after a scientist tried to patent the discovery.

The Onge had cures for many other ailments, deriving from their intimacy with the environment. They were blessed with "botanical and zoological knowledge which seems almost innate, and they know of properties in plants and animals of which we are quite unaware," wrote Cipriani. "They could tell me which trees flowered and when, because they knew this affected the whereabouts and the quality of the honey, and they knew which flowers and roots had medicinal properties." (Although his informants were men, the true repositories of botanical knowledge were the healers—the women. They could not, of course, treat introduced diseases such as tuberculosis, which require the medicines of outsiders.)

But with the removal of timber, the jungle that held this pharmacopeia was vanishing. A wilderness that to the Onge was made of a myriad trees, vines, and shrubs, each with its seasonally varying personality, its unique flowers and leaves and fruits, its aroma and taste and medicinal power, was to the outsider an amorphous, impenetrable mess. In the empty spaces left after logging, the newcomer often practiced agriculture, inserting the trees he'd brought along. Curiously, it was Cipriani who initiated the project of clearing the Onge's forests to make space for coconut groves: "In order to plant coconut trees for the benefit of the islanders I had even to do the planting myself, so utterly ignorant were they of agricultural methods; we even had to do all the clearing of the undergrowth in a patch of forest, and prepare the ground. The Onges watched curiously but were quite uninterested. Why should they care for a tree for ten years to get its nuts when the island and seas around are teeming with food for the taking?"

Quite so; and yet he persisted. Almost a hundred years earlier, the Great Andamanese, watching convicts at work in the fields, had been similarly unimpressed. According to Edward Man, they regarded agricultural labor as "a degrading occupation and fit only for such as have forfeited their freedom." The islanders' ways of acquiring food—hunting pigs, shooting fish, digging up tubers, plucking fruit, or wading into the lovely cool water to find clams—allowed for plenty of leisure.

"If a man has a painful dream he will often not venture out of the camp the following day, but will stay at home until the effect has worn off," Radcliffe-Brown had noted in passing. An odd detail, but of significance: which of us can take the day off because of a bad dream? By all accounts, the islanders' body and psyche were far better nourished than those of the modern Indian peasant. Man stated that the Great Andamanese had no concept of suicide and had coined a cumbersome compound word, *oyuntemar-toliganga*, upon observing it in outsiders.

But agriculture is synonymous with civilization, with enlightenment, with emancipation from bestial ignorance. So Indian officials had followed the British (and the Italian Cipriani) in striving tirelessly to teach cultivation to those unwilling students. The Jarawa contact trips often involved the planting of coconuts, which the *junglees* promptly dug up and ate.

Unfortunately, agriculture was a double-edged sword. Its advent allowed many more mouths to be fed and encouraged humans to multiply. But, as scientist and writer Jared Diamond relates, the balance between food and population was never quite right. Time and time again people became too numerous for their fields to support, and so they migrated and conquered until they reached the ends of the earth and nowhere left to go. Hunter-gatherers were the only peoples who knew how to live within their means.

Gaubolambe—Little Andaman—had been the Onge's universe. Their worldview and perhaps even their psyches were wrapped around its contours. Every beach, every stream, every redwood tree, every beehive they used to know and cherish.

Reading and Writing (2001)

Robert Wolff

When we talk about *education* we mean all that is involved in making us fit to live in civilized society. I remember the time I spent with people we would call primitive. They could not read because they had no use for it. That does not mean they were stupid. Being educated has little to do with being intelligent. In fact, the kind of intelligence people need to survive in the wild is usually destroyed by what we call education.

The Sng'oi do not have a written language. Their language is difficult to learn and is related to the Mong-Khmer language family, linguists say. It is not related to the many forms of Indonesian/Malay spoken by the people among whom they live today.

When I knew them, their villages—no more than settlements, really, of four or five shelters that might last the few years the People stayed in one place—were deep in the mountainous jungle in the center of Malaysia, near no roads (although there were one or two settlements close to a road). We had to walk along trails that were sometimes obvious but at other times required a guide. Certainly there were no telephones, no electricity, and no stores nearby—none of the amenities of life we take for granted.

I wanted to learn the language, so I carried a little notebook in which I would write down Sng'oi words, in the phonetic alphabet anthropologists use. Soon someone asked me about these mysterious scribbles in my notebook. I explained as well as I could that each scribble stood for a sound. By putting the sounds (technically speaking, phonemes) together, I could learn the words of their language. They thought that hilariously funny. They understood well enough that I wanted to learn their language, but why the scribbles? Could I not remember?

It was not the scribbles but my inability to memorize that was hilarious to them. I tried to explain, until a nice man who had not spoken before said he wanted to learn to write. Yes, others said, they did too.

I agreed to teach them. Tomorrow, they said. I knew that that did not mean the next day, but soon.

That evening I thought about how to teach. Each letter stands for a sound, combinations of letters stand for words ... I was not at all sure how to go about teaching basic reading and writing. I went to sleep with the comforting thought that they would probably forget about it anyway.

The next day came and went with nobody saying anything about learning to write. They were particularly busy that day because two young men had decided early in the morning to go hunting. The People hunted infrequently, using blowpipes and poisoned darts; more often they made elaborate traps to catch a wide variety of animals. The hunters came back in the afternoon with two little monkeys, which they gutted and put on a fire. The smell of singed hair was overpowering; there was little wind, and it stayed with us all night.

The next morning they said that they now wanted to learn to write. The entire village sat around—about a dozen people of all ages, including small children.

I wrote in the dirt: *A* stands for the sound *ah*. I scribbled again: *B* stands for the sound *buh*. We all repeated *ah* and *buh* a few times.

"Now put these two sounds together to make a word," I said.

They all sang out "*Ah-buh*." That did not make a word, however.

"Now the other way," I urged.

They said, in unison, "*Buh-ah*." Now faster: "*B-ah*." Hah! They recognized the word (*bah* means something like "mister"). Their name for me at that time was Bah Woo (*woo* was how they heard my last name).

They were delighted. They could write *ba*. They danced around, sang to each other, made jokes. "More," they said. "We want more." So very soon we learned other letters and put them together to make other words.

After the first hour I realized that they were far ahead of me. After all, they knew the language. I not only did not know their language but also was ignorant about how to teach the rudiments of writing. They corrected me when I said a word with the wrong emphasis or the wrong tone. Their language has a variety of strange and harsh consonants that don't exist in Malay, which on the whole is a very "soft" language. Malay/Indonesian is thought to be the mother language of, for instance, the Polynesian languages that have many vowel sounds and few consonants.

It was astonishing how easily and how well all of them memorized. It seemed that anything they heard and understood once, they knew. The first principle of education, we learn, is repetition. A teacher in our world repeats and repeats and repeats until it is drilled into a student's memory—that is how we think we have to teach. The Sng'oi, after seeing and hearing a letter once, knew it. No need to repeat anything.

Later I learned that this is not unusual. People whose minds have not been cluttered with endless facts have no trouble memorizing. That, by the way, is why oral history is probably as accurate as, if not more accurate than, what we call history.

After the second hour I became tired; they wanted to go on all day. The excitement never abated. No need to motivate them! The whole day was one hilarious adventure. They would give each other riddles in writing, then laugh. It was the best party they'd had in a long time, they said—better, even, than eating monkeys.

The next day someone asked, very earnestly, "Now what? What do we do with writing?"

What *could* they do with writing? There was nothing written in their language. No state-appointed committee had approved a written language. There were no newspapers, no books, no advertisements, no street names, no maps to read.

They decided they did not need writing after all.

If I wanted to play with those scribbles because otherwise I could not remember—they looked at each other with amazement and giggled again, thinking about my strange inability to memorize—that was fine for me, but they did not need anything to help them remember; they could remember without writing it down.

Learning letters had been fun but now they knew it was not really very useful.

They were right—in their world, it was not.

SECTION III

—

THE NATURE OF
CIVILIZATION

the white Raven

What is civilization? What is culture? Is it possible for a healthy race to be fathered by violence—in war or in the slaughter-house—and mothered by slaves, ignorant or parasitic? Where is the historian who traces the rise and fall of nations to the standing of women?

—Agnes Ryan (1952)

n Section Two, Chellis Glendinning suggests that civilization sprang forth after a long period in which its latent domesticating aspects slowly developed. A gradual, almost imperceptible, growth of specialization, or division of labor, may well have abetted this slippage toward a qualitatively new world of separation and control, as I have argued in *Elements of Refusal* (1988) and *Future Primitive* (1994). It seems evident that a struggle of contrary urges was involved; civilization never triumphs without a struggle.

In this section we are concerned with what civilization is, fundamentally. Does it have an inner logic? What is its core nature? At or near its center, its sheer authoritarianism must be recognized.

Michael Mann (1990) saw it this way:

In noncivilized societies escape from the social cage was possible. Authority was freely conferred, but recoverable; power, permanent and coercive, was unattainable.

A related fact is that every civilization in recorded history has routinely engaged in systematic and bloody warfare. It is hard to think of greater control, not to mention the grisly consequences, than that displayed by the institution of war.

Technology is another central locus of domination. Hans Jonas provides an apt description of this modern juggernaut, a cardinal fruit of the will to domesticate:

The danger of disaster attending the Baconian ideal of power over nature through scientific technology arises not so much from the shortcomings of its performance as from the magnitude of its success.

Civilization extends control over the natural and personal worlds, to ever greater lengths, in the direction of absolute manipulation.

In one of the entries in this section, the Unabomber cites the heightened powers of the modern order and the absence of fulfillment one experiences within its now global confines. Most of the other voices testify to other faces or facets: civilization as servitude and sacrifice, sickness, neurosis, psychological misery, frustration, repression, madness, frenzy, impoverishment, mass destruction, and self-destruction.

On the Aesthetic Education
of Man (1793)

Friedrich Schiller

C ivilization, far from setting us free, in fact creates some new
need with every new power it develops in us. The fetters of the
physical tighten ever more alarmingly, so that fear of losing what
we have stifles even the most burning impulse towards improvement, and the maxim of passive obedience passes for the supreme
wisdom of life.

Have I not perhaps been too hard on our age in the picture I
have just drawn? That is scarcely the reproach I anticipate. Rather
a different one: that I have tried to make it prove too much. Such
a portrait, you will tell me, does indeed resemble mankind as it is
today; but does it not also resemble any people caught up in the
process of civilization, since all of them, without exception, must
fall away from Nature by the abuse of Reason before they can
return to her by the use of Reason?...

It was civilization itself which inflicted this wound upon modern man. Once the increase of empirical knowledge, and more
exact modes of thought, made sharper divisions between the sciences inevitable, and once the increasingly complex machinery of
State necessitated a more rigorous separation of ranks and occu-
pations, then the inner unity of human nature was severed too,
and a disastrous conflict set its harmonious powers at variance.
The intuitive and the speculative understanding now withdrew in
hostility to take up positions in their respective fields, whose fron-
tiers they now began to guard with jealous mistrust; and with this
confining of our activity to a particular sphere we have given ourselves a master within, who not infrequently ends by suppressing
the rest of our potentialities. While in the one a riotous imagination ravages the hard-won fruits of the intellect, in another the
spirit of abstraction stifles the fire at which the heart should have
warmed itself and the imagination been kindled....

Thus, however much the world as a whole may benefit
through this fragmentary specialization of human powers, it cannot be denied that the individuals affected by it suffer under the

curse of this cosmic purpose. Athletic bodies can, it is true, be developed by gymnastic exercises; beauty only through the free and harmonious play of the limbs. In the same way the keying up of individual functions of the mind can indeed produce extraordinary human beings; but only the equal tempering of them all, happy and complete human beings. And in what kind of relation would we stand to either past or future ages, if the development of human nature were to make such sacrifice necessary? We would have been the serfs of mankind; for several millennia we would have done slaves' work for them, and our mutilated nature would bear impressed upon it the shameful marks of this servitude....

Man only plays when he is in the fullest sense of the word a human being, and he is only fully a human being when he plays.

Theory of Four Movements and General Destinies (1846)

Charles Fourier

Ａfter the catastrophe of 1793, illusions were swept away, and political science and moral philosophy were permanently stained and discredited. From that point, it became clear that all this acquired knowledge was useless. We had to look for the social good in some new science, and open new avenues for political genius. It was evident that neither the Philosophes or their rivals knew any remedies for social misery. Under either set of dogmas, the most shameful scourges would persist, poverty among them.

Such was the first consideration that led me to suspect the existence of a social science, as yet unknown, and prompted me to attempt to discover it. I was not scared off by my lack of learning; I set my sights on the honor of grasping what 25 centuries of scholars had not been able to figure out.

I was encouraged by numerous signs that reason had been led astray, and above all by the thought of the scourges that afflict social industry: indigence, the privation of work, the rewards of double-dealing, maritime piracy, commercial monopoly, the kidnapping of slaves ... so many misfortunes that they cannot be counted, and which give rise to the suspicion that civilized industry is nothing but a calamity, invented by God to punish humankind.

Given this premise, I assumed that this industry constituted some sort of disruption of the natural order; that it was carried out, perhaps, in a way that contradicted God's wishes; that the tenacity of so many scourges could be attributed to the absence of some arrangement willed by God, and unknown to our savants. Finally, I thought that if human societies are infected, as Montesquieu believed, "with a sickness of listlessness, with an inward vice, with a secret, hidden venom," the remedy might be found by avoiding the paths followed by our uncertain sciences that had failed to provide a remedy for so many centuries...."

CIVILIZATION AND ITS DISCONTENTS (1930)

SIGMUND FREUD

We come upon a contention which is so astonishing that we must dwell upon it. This contention holds that what we call our civilization is largely responsible for our misery, and that we should be much happier if we gave it up and returned to primitive conditions. I call this contention astonishing because, in whatever way we may define the concept of civilization, it is a certain fact that all the things with which we seek to protect ourselves against the threats that emanate from the sources of suffering are part of that very civilization.

How has it happened that so many people have come to take up this strange attitude of hostility to civilization? I believe that the basis of it was a deep and long-standing dissatisfaction with the then existing state of civilization and that on that basis a condemnation of it was built up, occasioned by certain specific historical events. I think I know what the last and the last but one of those occasions were. I am not learned enough to trace the chain of them far back enough in the history of the human species; but a factor of this kind hostile to civilization must already have been at work in the victory of Christendom over the heathen religions. For it was very closely related to the low estimation put upon earthly life by the Christian doctrine. The last but one of these occasions was when the progress of voyages of discovery led to contact with primitive peoples and races. In consequence of insufficient observation and a mistaken view of their manners and customs, they appeared to Europeans to be leading a simple, happy life with few wants, a life such as was unattainable by their visitors with their superior civilization. Later experience has corrected some of those judgements. In many cases the observers had wrongly attributed to the absence of complicated cultural demands what was in fact due to the bounty of nature and the ease with which the major human needs were satisfied. The last occasion is especially familiar to us. It arose when people came to know about the mechanism of the neuroses, which threaten to undermine the modicum of happiness enjoyed by civilized men. It was discovered that a person becomes neurotic because he cannot tolerate the amount of frustration which society imposes on him in the service of its cultural ideals, and it was inferred from this that the abolition or reduction of those demands would result in a return to possibilities of happiness.

There is also an added factor of disappointment. During the last few generations mankind has made an extraordinary advance in the natural sciences and in their technical application and has established his control over nature in a way never before imagined. The single steps of this advance are common knowledge and it is unnecessary to enumerate them. Men are proud of those achievements, and have a right to be. But they seem to have observed that this newly-won power over space and time, this subjugation of the forces of nature, which is the fulfillment of a longing that goes back thousands of years, has not increased the amount of pleasurable satisfaction which they may expect from life and has not made them feel happier....

If the development of civilization has such a far-reaching similarity to the development of the individual and if it employs the same methods, may we not be justified in reaching the diagnosis that, under the influence of cultural urges, some civilizations, or some epochs of civilization—possibly the whole of mankind—have become "neurotic"?

"Civilization and the Primitive" (1995)

John Landau

What is Primitive?

Primitive is trusting, trusting, trusting. It is naked. It is moving through language to ecstasy. The language of the body, of vocal cords, of dreams and vital ideas. The language of smells. Primitive is the naked foot touching the naked ground. Primitive is not without ideas, but ideas that hold together, embrace, and integrate the instincts. Primitive is dancing, the body moving on impulses. Primitive is letting go of confusion, embracing multiplicity. Primitive is getting lost in all you can be. Primitive is drifting, it is mastery of power expressed. It is bodies congregated in howling, painted packs. It is passion expressing itself in form; it is cunning grounding itself in stalking; it is cruelty creating itself through sensual ritual. Primitive is combat as love, pain as coming-closer, bewilderment as freefall into the ocean of being. It is wet, it is covered in mud, it is amniotic. Primitive is the gathering of plant and animal spirits; it is the hunting of mystery. It is difference bounding wild through fields of color, running free from ideals for the dangerous traps they are, fascinated all the same. Primitive is trusting the earth, the ground beneath one's feet.

What is Civilization?

Civilization is distrust, it is out-of-touch, it is pretense without play, pretending it is NOT. Civilization is only NOT; it makes sure it is NOT. It must have negation or it ceases to exist. Civilization is Either/Or. Civilization is "a place for everything and everything in its place" (but nowhere except in its place; while primitive is no place for any *thing*, for verb is all, as activity or rest). Civilization is the search for Extraterrestrial life, the desire to leave the earth. Civilization is disdain of the dirt, the soil, the mud. Civilization is being chained to the mind of Ideals, *any ideals*.

Civilization is perfection wreaking havoc on a squirming, squiggling, writhing, bumbling, blustering creature. It is the machine, fragmentation, the violation of integrity into coordinated parts. It is homelessness, exiled everywhere, therefore colonizing all it sees. It is life frustrated, frustrated, frustrated. It is manipulation. It is the belief that only the Good maintains the universe, for without it all would collapse; therefore, it is the demonizing of all it calls evil. It is the domination of paranoia, and therefore the paranoia of domination. Civilization is *not* fooling around, *not* blowing your top, *not* having a temper tantrum, *not* touching, *not* following your drift, *not* ease, *not* acting like those who are "lower" than you, *not* farting, *not* belching, *not* napping, *not* breathing, *not* crying, *not* resting.... It is a litany of "not's" (/knots). It has no substance, therefore it must overcome all it is not in order to prove to itself it exists. Civilization is Envy, it hates itself, the other side is greener, we must have it, the greed that comes from worthlessness, the desperate blotting out of the whole, therefore the feigning of superiority to save face, whoever saves the most face wins.

Eclipse of Reason (1947)

Max Horkheimer

Domination of nature involves domination of man. Each subject not only has to take part in the subjugation of external nature, human and nonhuman, but in order to do so must subjugate nature in himself. Domination becomes "internalized" for domination's sake. What is usually indicated as a goal—the happiness of the individual, health, and wealth—gains its significance exclusively from its functional potentiality. These terms designate favorable conditions for intellectual and material production. Therefore self-renunciation of the individual in industrialist society has no goal transcending industrialist society. Such abnegation brings about rationality with reference to means and irrationality with reference to human existence. Society and its institutions, no less than the individual himself, bear the mark of this discrepancy. Since the subjugation of nature, in and outside of man, goes on without a meaningful motive, nature is not really transcended or reconciled but merely repressed.

Resistance and revulsion arising from this repression of nature have beset civilization from its beginnings, in the form of social rebellions—as in the spontaneous peasant insurrections of the sixteenth century or the cleverly staged race riots of our own day—as well as in the form of individual crime and mental derangement. Typical of our present era is the manipulation of this revolt by the prevailing forces of civilization itself, the use of the revolt as a means of perpetuating the very conditions by which it is stirred up and against which it is directed. Civilization as rationalized irrationality integrates the revolt of nature as another means or instrument....

The story of the boy who looked up at the sky and asked, "Daddy, what is the moon supposed to advertise?" is an allegory of what has happened to the relation between man and nature in the era of formalized reason. On the one hand, nature has been stripped of all intrinsic value or meaning. On the other, man has been stripped of all aims except self-preservation. He tries to transform everything within reach into a means to that end. Every word or sentence that hints of relations other than pragmatic is suspect. When a man is asked to admire a thing, to respect a feeling or attitude, to love a person for his own sake, he smells sentimentality and suspects that someone is pulling his leg or trying to sell him something. Though people may not ask what the moon is supposed to advertise, they tend to think of it in terms of ballistics or aerial mileage.

The complete transformation of the world into a world of means rather than of ends is itself the consequence of the historical development of the methods of production. As material production and social organization grow more complicated and reified, recognition of means as such becomes increasingly difficult, since they assume the appearance of autonomous entities. As long as the means of production are primitive, the forms of social organization are primitive....

Speculative thought, from the economic point of view, was doubtless a luxury that, in a society based on group domination only a class of people exempt from hard labor could afford. The intellectuals, for whom Plato and Aristotle were the first great European spokesmen, owe their very existence, and their leisure to indulge in speculation, to the system of domination from which they try to emancipate themselves intellectually. The vestiges of this paradoxical situation can be discovered in various systems of thought. Today—and this is certainly progress—the masses know that such freedom for contemplation crops up only occasionally. It was always a privilege of certain groups, which automatically built up an ideology hypostatizing their privilege as a human virtue; thus it served actual ideological purposes, glorifying those exempt from manual labor. Hence the distrust aroused by the group. In our era the intellectual is, indeed, not exempt from the pressure that the economy exerts upon him to satisfy the ever-changing demands of reality. Consequently, mediation, which looked to eternity, is superseded by pragmatic intelligence, which looks to the next moment. Instead of losing its character as a privilege, speculative thought is altogether liquidated—and this can hardly be called progress. It is true that in this process nature has lost its awesomeness, its *qualitates occultae*, but, completely deprived of the chance to speak through the minds of men even in the distorted language of these privileged groups, nature seems to be taking its revenge.

Modern insensitivity to nature is indeed only a variation of the pragmatic attitude that is typical of Western civilization as a whole. The forms are different. The early trapper saw in the prairies and mountains only the prospects of good hunting; the modern businessman sees in the landscape an opportunity for the display of cigarette posters. The fate of animals in our world is symbolized by an item printed in newspapers of a few years ago. It reported that landings of planes in Africa were often hampered by herds of elephants and other beasts. Animals are here considered simply as obstructors of traffic. This mentality of man as the master can be traced back to the first chapters of Genesis. The few precepts in favor of animals that we encounter in the Bible have been interpreted by the most outstanding religious thinkers, Paul, Thomas Aquinas, and Luther, as pertaining only to the moral education of man, and in no wise to any obligation of man toward other creatures. Only man's soul can be saved; animals have but the right to suffer. "Some men and women," wrote a British churchman a few years ago, "suffer and die for the life, the welfare, the happiness of others. This law is continually seen in operation. The supreme example of it was shown to the world (I write with reverence) on Calvary. Why should animals be exempted from the operation of this law or principle?" Pope Pius IX did not permit a society for the prevention of cruelty to animals to be founded in Rome because, as he declared, theology teaches that man owes no duty to any animal. National Socialism, it is true, boasted of its protection of animals, but only in order to humiliate more deeply those "inferior races" whom they treated as mere nature....

DAWN AND DECLINE (1961)

MAX HORKHEIMER

n the Circus: Through the image of the elephant in the circus, man's technological superiority becomes conscious of itself. With whip and iron hooks, the ponderous animal is brought in. On command, it raises its right, its left foot, its trunk, describes a circle, lies down laboriously and finally, as the whip is being cracked, it stands on two legs which can barely support the heavy body.

For many hundreds of years, that's what the elephant has had to do to please people. But one should say nothing against the circus or the act in the ring. It is no more foreign, no more inappropriate, probably more suitable to the animal than the slave labor for whose sake it entered human history. In the arena, where the elephant looks like the image of eternal wisdom as it confronts the stupidity of the spectators and where, among fools, it makes a few foolish gestures for the sake of peace and quiet, the objective unreason of the compulsory service which serves the rational purpose of the Indian timber market still reveals itself. That men depend on such labor to then be obliged to subject themselves to it as well is ultimately their own disgrace. The enslavement of the animal as the mediation of their existence through work that goes against their own and alien nature has the result that that existence is as external to them as the circus act is to the animal. Rousseau had an intimation of this when he wrote his prize-winning essays. Civilization as stultification.

"Was Civilization a Mistake?" (1997)

Richard Heinberg

aving been chosen—whether as devil's advocate or sacrificial lamb, I am not sure—to lead off this discussion on the question, "Was Civilization a Mistake?", I would like to offer some preliminary thoughts.

From the viewpoint of any non-civilized person, this consideration would appear to be steeped in irony. Here we are, after all, some of the most civilized people on the planet, discussing in the most civilized way imaginable whether civilization itself might be an error. Most of our fellow civilians would likely find our discussion, in addition to being ironic, also disturbing and pointless: after all, what person who has grown up with cars, electricity, and television would relish the idea of living without a house, and of surviving only on wild foods?

Nevertheless, despite the possibility that at least some of our remarks may be ironic, disturbing, and pointless, here we are. Why? I can only speak for myself. In my own intellectual development I have found that a critique of civilization is virtually inescapable for two reasons.

The first has to do with certain deeply disturbing trends in the modern world. We are, it seems, killing the planet. Revisionist "wise use" advocates tell us there is nothing to worry about; dangers to the environment, they say, have been wildly exaggerated. To me this is the most blatant form of wishful thinking. By most estimates, the oceans are dying, the human population is expanding far beyond the long-term carrying capacity of the land, the ozone layer is disappearing, and the global climate is showing worrisome signs of instability. Unless drastic steps are taken, in 50 years the vast majority of the world's population will likely be existing in conditions such that the lifestyle of virtually any undisturbed primitive tribe would be paradise by comparison.

Now, it can be argued that civilization *per se* is not at fault, that the problems we face have to do with unique economic and historical circumstances. But we should at least consider the possibility that our modern industrial system represents the flowering of tendencies that go back quite far. This, at any rate, is the implication of recent assessments of the ecological ruin left in the wake of the Roman, Mesopotamian, Chinese and other prior civilizations. Are we perhaps repeating their errors on a gargantuan scale?

If my first reason for criticizing civilization has to do with its effects on the environment, the second has to do with its impact on human beings. As civilized people, we are also domesticated. We are to primitive peoples as cows and sheep are to bears and eagles. On the rental property where I live in California my landlord keeps two white domesticated ducks. These ducks have been bred to have wings so small as to prevent them from flying. This is a convenience for their keepers, but compared to wild ducks these are pitiful creatures.

Many primal peoples tend to view us as pitiful creatures, too—though powerful and dangerous because of our technology and sheer numbers. They regard civilization as a sort of social disease. We civilized people appear to act as though we were addicted to a powerful drug—a drug that comes in the forms of money, factory-made goods, oil, and electricity. We are helpless without this drug, so we have come to see any threat to its supply as a threat to our very existence. Therefore we are easily manipulated—by desire (for more) or fear (that what we have will be taken away)—and powerful commercial and political interests have learned to orchestrate our desires and fears in order to achieve their own purposes of profit and control. If told that the production of our drug involves slavery, stealing, and murder, or the ecological equivalents, we try to ignore the news so as not to have to face an intolerable double bind.

Since our present civilization is patently ecologically unsustainable in its present form, it follows that our descendants will be living very differently in a few decades, whether their new way of life arises by conscious choice or by default. If humankind is to choose its path deliberately, I believe that our deliberations should include a critique of civilization itself, such as we are undertaking here. The question implicit in such a critique is, *What we have done poorly or thoughtlessly in the past that we can do better now?* It is in this constructive spirit that I offer the comments that follow.

What is Primitivism?

The image of a lost Golden Age of freedom and innocence is at the heart of all the world's religions, is one of the most powerful themes in the history of human thought, and is the earliest and most characteristic expression of *primitivism*—the perennial belief in the necessity of a return to origins.

As a philosophical idea, primitivism has had as its proponents: Lao Tze, Rousseau, and Thoreau, as well as most of the pre-Socratics, the medieval Jewish and Christian theologians, and 19th- and 20th-century anarchist social theorists, all of whom argued (on different bases and in different ways) the superiority of a simple life close to nature. More recently, many anthropologists have expressed admiration for the spiritual and material advantages of the ways of life of the world's most "primitive" societies—the surviving gathering-and-hunting peoples who now make up less than one hundredth of one percent of the world's population.

Meanwhile, as civilization approaches a crisis precipitated by overpopulation and the destruction of the ecological integrity of the planet, primitivism has enjoyed a popular resurgence, by way of increasing interest in shamanism, tribal customs, herbalism, radical environmentalism, and natural foods. There is a widespread (though by no means universally shared) sentiment that civilization has gone too far in its domination of nature, and that in order to survive—or, at least, to live with satisfaction—we must regain some of the spontaneity and naturalness of our early ancestors.

What is Civilization?

There are many possible definitions of the word *civilization*. Its derivation—from *civis*, "town" or "city"—suggests that a minimum definition would be "urban culture." Civilization also seems to imply writing, division of labor, agriculture, organized warfare, growth of population, and social stratification.

Yet the latest evidence calls into question the idea that these traits always go together. For example, Elizabeth Stone and Paul Zimansky's assessment of power relations in the Mesopotamian city of Maskan-shapir (published in the April 1995 *Scientific American*) suggests that urban culture need not imply class divisions. Their findings seem to show that civilization in its earliest phase was free of these. Still, for the most part the history of civilization in the Near East, the Far East, and Central America, is also the history of kingship, slavery, conquest, agriculture, overpopulation, and environmental ruin. And these traits continue in civilization's most recent phases—the industrial state and the global market—though now the state itself takes the place of the king, and slavery becomes wage labor and *de facto* colonialism administered through multinational corporations. Meanwhile, the mechanization of production (which began with agriculture) is overtaking nearly every avenue of human creativity, population is skyrocketing and organized warfare is resulting in unprecedented levels of bloodshed....

Wild Self/Domesticated Self

People are shaped from birth by their cultural surroundings and by their interactions with the people closest to them. Civilization manipulates these primary relationships in such a way as to domesticate the infant—that is, so as to accustom it to life in a social structure one step removed from nature. The actual process of domestication is describable as follows, using terms borrowed from the object-relations school of psychology.

The infant lives entirely in the present moment in a state of pure lust and guilelessness, deeply bonded with her mother. But as she grows, she discovers that her mother is a separate entity with her own priorities and limits. The infant's experience of relationship changes from one of spontaneous trust to one that is suffused with need and longing. This creates a gap between Self and Other in the consciousness of the child, who tries to fill this deepening rift with *transitional objects*—initially, perhaps a teddy bear; later, additions and beliefs that serve to fill the psychic gap and thus provide a sense of security. It is the powerful human need for transitional objects that drives individuals in their search for property and power, and that generates bureaucracies and technologies as people pool their efforts.

This process does not occur in the same way in the case of primitive childbearing, where the infant is treated with indulgence, is in constant physical contact with a caregiver throughout infancy, and later undergoes rites of passage. In primal cultures the need for transitional objects appears to be minimized. Anthropological and psychological research converge to suggest that many of civilized people's emotional ills come from our culture's abandonment of natural childrearing methods and initiatory rites and its systematic substitution of alienating pedagogical practices from crib through university.

Health: Natural or Artificial?

In terms of health and quality of life, civilization has been a mitigated disaster. S. Boyd Eaton, M.D., *et al.*, argued in *The*

Paleolithic Prescription (1988) that pre-agricultural peoples enjoyed a generally healthy way of life, and that cancer, heart disease, strokes, diabetes, emphysema, hypertension, and cirrhosis—which together lead to 75 percent of all mortality in industrialized nations—are caused by our civilized lifestyles. In terms of diet and exercise, preagricultural lifestyles showed a clear superiority to those of agricultural and civilized peoples.

Much-vaunted increases in longevity in civilized populations have resulted not so much from wonder drugs, as merely from better sanitation—a corrective for conditions created by the overcrowding of cities; and from reductions in infant mortality. It is true that many lives have been spared by modern antibiotics. Yet antibiotics also appear responsible for the evolution of resistant strains of microbes, which health officials now fear could produce unprecedented epidemics in the next century.

The ancient practice of herbalism, evidence of which dates back at least 60,000 years, is practiced in instinctive fashion by all higher animals. Herbal knowledge formed the basis of modern medicine and remains in many ways superior to it. In countless instances, modern synthetic drugs have replaced herbs not because they are safer or more effective, but because they are more profitable to manufacture.

Other forms of "natural" healing—massage, the "placebo effect," the use of meditation and visualization—are also being shown effective. Medical doctors Bernie Siegel and Deepak Chopra are critical of mechanized medicine and say that the future of the healing profession lies in the direction of attitudinal and natural therapies.

Spirituality: Raw or Cooked?

Spirituality means different things to different people—humility before a higher power or powers; compassion for the suffering of others; obedience to a lineage or tradition; a felt connection with the Earth or with Nature; evolution toward "higher" states of consciousness; or the mystical experience of oneness with all life or with God. With regard to each of these fundamental ways of defining or experiencing the sacred, spontaneous spirituality seems to become regimented, dogmatized, even militarized, with the growth of civilization. While some of the founders of world religions were intuitive primitivists (Jesus, Lao Tze, the Buddha), their followers have often fostered the growth of dominance hierarchies.

The picture is not always simple, though. The thoroughly civilized Roman Catholic Church produced two of the West's great primitivists—St. Francis and St. Clair; while the neo-shamanic, vegetarian, and herbalist movements of early 20th-century Germany attracted arch-authoritarians Heinrich Himmler and Adolph Hitler. Of course, Nazism's militarism and rigid dominator organization were completely alien to primitive life, while St. Francis' and St. Clair's voluntary poverty and treatment of animals as sacred were reminiscent of the lifestyle and worldview of most gathering-and-hunting peoples. If Nazism was atavistic, it was only highly selectively so....

Economics: Free or Affordable?

At its base, economics is about how people relate with the land and with one another in the process of fulfilling their material wants and needs. In the most primitive societies, these relations are direct and straightforward. Land, shelter, and food are free. Everything is shared, there are no rich people or poor people, and happiness has little to do with accumulating material possessions. The primitive lives in relative abundance (all needs and wants are easily met) and has plenty of leisure time.

Civilization, in contrast, straddles two economic pillars—technological innovation and the marketplace. "Technology" here includes everything from the plow to the nuclear reactor—all are means to more efficiently extract energy and resources from nature. But efficiency implies the reification of time, and so civilization always brings with it a preoccupation with past and future; eventually the present moment nearly vanishes from view. The elevation of efficiency over other human values is epitomized in the factory—the automated workplace—in which the worker becomes merely an appendage of the machine, a slave to clocks and wages.

The market is civilization's means of equating dissimilar things through a medium of exchange. As we grow accustomed to valuing everything according to money, we tend to lose a sense of the uniqueness of things. What, after all, is an animal worth, or a mountain, or a redwood tree, or an hour of human life? The market gives us a numerical answer based on scarcity and demand. To the degree that we believe that such values have meaning, we live in a world that is desacralized and desensitized, without heart or spirit.

We can get some idea of ways out of our ecologically ruinous, humanly deadening economic cage by examining not only primitive lifestyles, but the proposals of economist E.F. Schumacher, the experiences of people in utopian communities in which technology and money are marginalized, and the lives of individuals who have adopted an attitude of voluntary simplicity.

Government: Bottom Up or Top Down?

In the most primitive human societies there are no leaders, bosses, politics, laws, crime, or taxes. There is often little division of labor between women and men, and where such division exists both gender's contributions are often valued more or less equally. Probably as a result, many foraging peoples are relatively peaceful. Anthropologist Richard Lee found that the !Kung [Bushmen of southern Africa] hate fighting, and think anybody who fought would be stupid.

With agriculture usually come division of labor, increased sexual inequality, and the beginnings of social hierarchy. Priests, kings, and organized impersonal warfare all seem to come together in one package. Eventually, laws and borders define the creation of the fully fledged state. The state as a focus of coercion and violence has reached its culmination in the 19th and 20th centuries in colonialism, fascism, and Stalinism. Even the democratic industrial state functions essentially as an instrument of multinational corporate-style colonial oppression and domestic enslavement, its citizens merely being given the choice between selected professional bureaucrats representing political parties with slightly varying agendas for the advancement of corporate power.

Beginning with William Godwin in the early 19th century, anarchist social philosophers have offered a critical counterpoint to the increasingly radical statism of most of the world's civilized political leaders. The core idea of anarchism is that human beings are fundamentally sociable; left to themselves, they tend to cooperate to their mutual benefit. There will always be exceptions, but these are best dealt with informally and on an individual basis. Many anarchists cite the Athenian polis, the "sections" in Paris during the French Revolution, the New England town meetings of the 18th century, the popular assemblies in Barcelona in the late 1930s, and the Paris general strike of 1968 as positive examples of anarchy in action. They point to the possibility of a kind of social ecology, in which diversity and spontaneity are permitted to flourish unhindered both in human affairs and in Nature....

Civilization and Nature

Civilized people are accustomed to an anthropocentric view of the world. Our interest in the environment is utilitarian: it is of value because it is of use (or potential use) to human beings—if only as a place for camping and recreation.

Primitive peoples, in contrast, tended to see nature as intrinsically meaningful. In many cultures prohibitions surrounded the overhunting of animals or the felling of trees. The aboriginal peoples of Australia believed that their primary purpose in the cosmic scheme of things was to take care of the land, which meant performing ceremonies for the periodic renewal of plant and animal species, and of the landscape itself. The difference in effects between the anthropocentric and ecocentric worldviews is incalculable. At present, we human beings—while considering ourselves the most intelligent species on the planet—are engaged in the most unintelligent enterprise imaginable: the destruction of our own natural life-support system. We need here only mention matters such as the standard treatment of factory-farmed domesticated food animals, the destruction of soils, the pollution of air and water, and the extinctions of wild species, as these horrors are well documented. It seems unlikely that these could ever have arisen but for an entrenched and ever-deepening trend of thinking that separates humanity from its natural context and denies inherent worth to non-human nature....

HERE: A SMALL HISTORY OF A
MINING TOWN IN THE AMERICAN SOUTHWEST:
WARREN/BISBEE AZ (1985)

BARBARA MOR

From the top rim downward, sheer cascades of colors: mauve, gold, rust, purple, pink, silver, blue, incandescent turquoise. Streaks of orange, streaks of fire, yellow streaks of toxic arsenic. Radioactive greens of lime and fungus. Each color spilled over the others, in corroded terraced levels, channeled by erosion, avalanched by rain, crusted, broken open, merged; each geologic texture, as though alive, crawling over the variegated lumps and rubble of the earthly flesh that came before. And over this, the solid spills of individual rocks, orthorhombic crystals, dredging gears, rusted-out elbows, coils of wire, buttons, nails, hair curlers, stray lead bullets, all runneling, flowing in geological slow motion one over another down to the center of the hole. Deep, deep down. The entire technological history of the Pit was thus laid bare to any observer, in concentric layer after layer, vast polychronic slide upon slide, sloping down from the first simple surface diggings, by hand and stick, of the precivilized beings, immemorial years ago, downward through sediments of beauty, sediments of grief, sediments of nothing very important or useful, sediments of historical overthrow, one solid layer of crushed bones; and then further down the notable sediments of great wealth and petrochemical power, sediments of capital gains, sediments of wrapping paper, one solid sedimentary level each of glass, electrical conducting alloy, and stockpiled war material....

When first civilized beings arrived in the region, in wagons and on horseback, they found indigenous naked creatures squatting in the dirt, digging up small geological objects with bare hands and pointed sticks. Thus the Pit's origin, in what was then only a slight depression of the earth, a sandy little dip. The objects sought by the natives were simple gem stones, turquoise, azurite, malachite, roughly polished and used ornamentally. In the childlike mental grasp of these early beings there was no concept of serious mining, a factor which helped account for their elimination....

With settlement, mining began in earnest. Men came from everywhere, attracted by the adventure.... The Pit grew into the largest intentional hole on the earth's surface. Innovators, in the early years, introduced various subterranean approaches to the extraction of ores from earth, but the straightforward digging of a hole, deeper and deeper into the ground, always seemed the most expeditious method for this terrain. Several hundred males, equipped with picks and shovels, simply began digging; as the hole grew, timbers were used to shore up the higher levels of dirt, and these shorings congealed into a circumference of terraces. Later, tracks were built, for railcars, and mules brought in to haul them; deeper down, around and around; the workforce grew from a few hundred to thousands. When groundwater seeped into working levels, as it did more often at greater depths, giant sponges were brought in to soak up the intruding fluids. When water flooded in, violently, without warning, drowning hundreds of workers, and burying hundreds more beneath tons of collapsed earth, the sponges were hooked up behind the mule-drawn railcars and dragged down and around the wet circumference, gradually soaking up the waters. The sponges were periodically wrung out by giant rollers (similar to old-fashioned washing machine wringers), into the railcars, and the water hauled by more patient mules around and upward to the dry surface; the method was as simple as it was efficient. Drowned bodies were usually soaked up also, lodging in the sponge holes, and removed by the same process; if not, the bones were extracted by shovels from dried sediments at a later date. At great depths, the sponges were working constantly, and pipe systems were eventually installed to transport the fluids, which at some point became quite valuable.

The first substances ... ordinary gold, silver, copper.... Gears. When the legendary gear ran out, no problem. It was followed in quick succession by equally enormous outpourings of nails, ball-bearings, screws, the aforementioned lead bullets, and office paper clips. Car windshields and batteries alternated with hand grenades and gasmasks. At one point cigarette lighters flowed out at the rate of 1,523 per minute in precise alternation with 1,523 cans of lighter fluid; this was troublesome when they ignited each other (due to worker error) and the entire level erupted in a blazing inferno of metallic flames reaching almost to the Pit's rim. It burned for three nights and three days. Untold numbers of workers were lost, along with the tragic destruction of 8,632,948 cans of lighter fluid and 8,632,948 cigarette lighters.

When the smoke cleared, the Pit revealed walls of char and molten rivulets; a season's rain was needed to wash down the blackness.... This was a difficult time. The fire had sealed over, cauterized the hundreds of thousands of productive little holes from which such great abundance had recently poured. The earth was streaked with hardness, a surface meld of alloys from so many stray bullets, gears, paper clips, cigarette lighters, etc. Newly recruited workteams went down into the Pit to dig with pickaxes and sledgehammers. As they shattered, uprooted, peeled back this fused metalloid carapace from large scarred flanks and thighs of the damaged inner hole, they uncovered raw blotches of more disgusting things: layers upon layers of half-corroded used sanitary napkins, douche bags, enema hoses and syringes, broken rotten teeth, and the overwhelming stench of something dead. As before, but with increased efficiency, these nauseating items were gathered into large heaps and disposed of immediately. (The death smell of course lingered until the production of aerosol cans.) And then around the edges of these picked scabs, as it were, from around the nocturnal fringes of such terrible scars and unmentionable uses, something new began to ooze. A gooey substance, pellucid green in color. Shyly at first, and then with increasing ebullience, it crept and flowed and jiggled over the lower Pit surface, covering over the recent devastation like an innocent vegetation or spring grass. Something about it invited tasting. Several workers vomited at the thought. But then one, then another, then another and another, bent over and dipped a delicate forefinger in the happy green substance overwhelming their rubber boots, now, to the knees; and tasted. And found it good.

Lime gelatin.

It was mid-century. A difficult period had been experienced, but a challenging one....

Then the deep well of Ink turned red ... viscid red; when managers went down to test it, their pens clogged, or the words they wrote all looked like death scrawls. And the air was saturated with a thick odor, unlike all others; Dobermans guarding the town's five mansions howled day and night, an incessant whining that became invisible, inaudible as the voice of everything and everyone.

It soaked the rolling hills, pushing gelatinous clots to the surface of coarse dirt. Cracks in the sidewalks filled with red, cement and dust lots permeated with red stains. Red flowing gutters, the flood ditches awash with red. Citizens sprayed driveways and garages with lawn hoses, trying to prevent the disfigurement of cars. But the blood seeped everywhere, and the arroyos surrounding the town were as open veins.... Trees and bushes began to suck up the red fluid from their roots, as did the porous walls of the great houses. Within five days it had entered the town's water supply. A mine manager, sprinkling his front lawn in the summer evening of a dry day, turned his hose on a little garden patch of cacti, oleander and budding agave. The black-flecked water spraying out suddenly turned to rust, clogged, and then with explosive force began pouring a stream of pure blood. With the mental control of an executive he continued spraying plants, a perfect, silent adjustment to this final change, as the red viscosity covered his lawn and garden, his gray pants and canvas shoes, with spatters, globs and blotches of an irrepressible bloody dew....

With the appearance of blood in the water pipes, the town's small hospital staff fled, citing a conflict of interest. And with great regret, the Mine Museum closed. The few visitors who did come, after the news got out, were reluctant to park their cars and then walk 25 yards through sticky, scarlet clots to the Museum entrance. In high winds the clots moved, like tumbleweeds of blood. A bridge was built, to accommodate these tourists; but the gluey blobs continued to mass and ooze on the threshold and pile up against the Museum windows. Beings not used to it found this offensive. With cessation of the tourist trade, the town's residents also realized the market shelves had grown quite bare; no food had been delivered for weeks. At this point the Town Council voted to become officially self-sufficient. Indeed, with blood flowing freely through all the town's plumbing, from home faucets, from drinking fountains in banks and gas stations and even the dusty park, most beings had become quite satisfied with this diet....

Some worried about the town's isolation. Incoming roads were silent, mail to and from the outside world ceased. Rarely, a small private airplane flew over in the late afternoon, attempting colored postcard shots of the Pit; but these were sensationalists, who never landed, who were not interested in the ordinary, daily life of the town. In a lonely show of spirit, the postmaster and two postal workers devised a combination stamp and postmark for the local mail. Using sponges soaked in blood from the bathroom tap, they dipped and affixed their bloody thumbprints to the top righthand corner of every envelope dropped through their slots. Unfortunately, town residents had little to say to each other, by mail or otherwise; the post office closed. Eventually the postal workers were reemployed as trash burners at the dumpsite east of town. They now shovel great white and red heaps of paper products into the flames, blood-saturated napkins, tablecloths and medical gowns that smoulder and crackle thickly throughout the night, releasing a stench similar to burning corpse flesh; which, of course, everyone has grown used to....

From its experience, the town has learned something profound about the nature of its own will, as of the mysterious hidden resources of the Earth. Deeper and deeper, as it had descended into the dark downward and abysm not solely of time but of its own evolution, what it had dug with its historic fingers from this soiled Hole, so to speak, was an implacable knowledge others could profit from, if others would: That the inexorable becomes the simply inextricable, and thus the normal; and vice versa. If only beings strive to make it so. The question of whose blood is never raised. Nor, if the hole extends through the globe to China, could it be Communist blood? Intellectual quibbling is extraneous to the town's experience of itself.

Some beings from the outside have called the Pit a Living Wound, citing the bloodflow as a strong proof. But morbid and negative metaphors do not make the world turn; as gears do, for example, or ball bearings. Or as now, the mining of blood.

Toward a History of Needs (1978)

Ivan Illich

𝔐 odernized poverty appears when the intensity of market dependence reaches a certain threshold. Subjectively, it is the experience of frustrating affluence which occurs in persons mutilated by their overwhelming reliance on the riches of industrial productivity. Simply, it deprives those affected by it of their freedom and power to act autonomously, to live creatively; it confines them to survival through being plugged into market relations. And precisely because this new impotence is so deeply experienced, it is with difficulty expressed. We are the witnesses of a barely perceptible transformation in ordinary language by which verbs that formerly designated satisfying actions are replaced by nouns that denote packages designed for passive consumption only: for example, "to learn" becomes "acquisition of credits." A profound change in individual and social self-images is here reflected. And the layman is not the only one who has difficulty in accurately describing what he experiences. The professional economist is unable to recognize the poverty his conventional instruments fail to uncover. Nevertheless, the new mutant of impoverishment continues to spread. The peculiarly modern inability to use personal endowments, communal life, and environmental resources in an autonomous way infects every aspect of life where a professionally engineered commodity has succeeded in replacing a culturally shaped use-value. The opportunity to experience personal and social satisfaction outside the market is thus destroyed. I am poor, for instance, when the use-value of my feet is lost because I live in Los Angeles on the 35th floor.

This new impotence-producing poverty must not be confused with the widening gap between the consumption of rich and poor in a world where basic needs are increasingly shaped by industrial commodities. That gap is the form traditional poverty assumes in an industrial society, and the conventional terms of class struggle appropriately reveal and reduce it. I further distinguish modernized poverty from the burdensome price exacted by the externalities which increased levels of production spew into the environment. It is clear that these kinds of pollution, stress, and taxation are unequally imposed. Correspondingly, defenses against such depredations are unequally distributed. But like the new gaps in access, such inequities in social costs are aspects of industrialized poverty for which economic indicators and objective verification can be found. Such is not true for the industrialized impotence which affects both rich and poor. Where this kind of poverty reigns, life without addictive access to commodities is rendered either impossible or criminal. Making do without consumption becomes impossible, not just for the average consumer but even for the poor. All forms of welfare, from affirmative action to environmental action, are of no help. The liberty to design and craft one's own distinctive dwelling is abolished in favor of the bureaucratic provision of standardized housing, as in the United States, Cuba or Sweden. The organization of employment, skills, building resources, rules, and credit favor shelter as a commodity rather than as an activity. Whether the product is provided by an entrepreneur or an apparatchik, the effective result is the same: citizen impotence, our specifically modern experience of poverty.

Wherever the shadow of economic growth touches us, we are left useless unless employed on a job or engaged in consumption; the attempt to build a house or set a bone outside the control of certified specialists appears as anarchic conceit. We lose sight of our resources, lose control over the environmental conditions which make these resources applicable, lose taste for self-reliant coping with challenges from without and anxiety from within. Take childbirth in Mexico today: delivery without professional care has become unthinkable for those women whose husbands are regularly employed and therefore have access to social services, no matter how marginal or tenuous. They move in circles where the production of babies faithfully reflects the patterns of industrial outputs. Yet their sisters in the slums of the poor or the villages of the isolated still feel quite competent to give birth on their own mats, unaware that they face a modern indictment of criminal neglect toward their infants. But as professionally engineered delivery models reach these independent women, the desire, competence, and conditions for autonomous behavior are being destroyed.

For advanced industrial society, the modernization of poverty means that people are helpless to recognize evidence unless it has been certified by a professional, be he a television weather commentator or an educator; that organic discomfort becomes intolerably threatening unless it has been medicalized into dependence on a therapist; that neighbors and friends are lost unless vehicles bridge the separating distance (created by the vehicles in the first place). In short, most of the time we find ourselves out of touch with our world, out of sight of those for whom we work, out of tune with what we feel.

Modernity and the Holocaust (1989)

Zygmunt Bauman

The meaning of the civilizing process

The etiological myth deeply entrenched in the self-consciousness of our Western society is the morally elevating story of humanity emerging from pre-social barbarity. This myth lent stimulus and popularity to, and in turn was given a learned and sophisticated support by, quite a few influential sociological theories and historical narratives....

In view of this myth, long ago ossified into the common sense of our era, the Holocaust can only be understood as the failure of civilization (i.e., of human purposive, reason-guided activity) to contain the morbid natural predilections of whatever has been left of nature in man. Obviously, the Hobbesian world has not been fully chained, the Hobbesian problem has not been fully resolved. In other words, we do not have as yet enough civilization. The unfinished civilizing process is yet to be brought to its conclusion. If the lesson of mass murder does teach us anything it is that the prevention of similar hiccups of barbarism evidently requires still more civilizing efforts. There is nothing in this lesson to cast doubt on the future effectiveness of such efforts and their ultimate results. We certainly move in the right direction; perhaps we do not move fast enough.

As its full picture emerges from historical research, so does an alternative, and possible more credible, interpretation of the Holocaust as an event which disclosed the weakness and fragility of human nature (of the abhorrence of murder, disinclination to violence, fear of guilty conscience and of responsibility for immoral behaviour) when confronted with the matter-of-fact efficiency of the most cherished among the products of civilization; its technology, its rational criteria of choice, its tendency to subordinate thought and action to the pragmatics of economy and effectiveness. The Hobbesian world of the Holocaust did not surface from its too-shallow grave, resurrected by the tumult of irrational emotions. It arrived (in a formidable shape Hobbes would certainly disown) in a factory-produced vehicle, wielding weapons only the most advanced science could supply, and following an itinerary designed by scientifically managed organization. Modern civilization was not the Holocaust's *sufficient* condition; it was, however, most certainly its *necessary* condition. Without it, the Holocaust would be unthinkable. It was the rational world of modern civilization that made the Holocaust thinkable. The Nazi mass murder of the European Jewry was not only the technological achievement of an industrial society, but also the organizational achievement of a bureaucratic society....

The most shattering of lessons deriving from the analysis of the 'twisted road to Auschwitz' is that—in the last resort—*the choice of physical extermination as the right means to the task of* Entfernung *was a product of routine bureaucratic procedures*: means-end calculus, budget balancing, universal rule application. To make the point sharper still the choice was an effect of the earnest effort to find rational solutions to successive 'problems', as they arose in the changing circumstances. It was also affected by the widely described bureaucratic tendency to goal-displacement—an affliction as normal in all bureaucracies as their routines. The very presence of functionaries charged with their specific tasks led to further initiatives and a continuous expansion of original purposes. Once again, expertise demonstrated its self-propelling capacity, its proclivity to expand and enrich the target which supplied its *raison d'etre*.

> The mere existence of a corpus of Jewish experts created a certain bureaucratic momentum behind Nazi Jewish policy. Even when deportations and mass murder were already under way, decrees appeared in 1942 prohibiting German Jews from having pets, getting their hair cut by Aryan barbers, or receiving the Reich sport badge! It did not require orders from above, merely the existence of the job itself, to ensure that the Jewish experts kept up the flow of discriminating measures.

At no point of its long and tortuous execution did the Holocaust come in conflict with the principles of rationality. The 'Final Solution' did not clash at any stage with the rational pursuit of efficient, optimal goal-implementation. On the contrary, *it arose out of a genuinely rational concern, and it was generated by bureaucracy true to its form and purpose*. We know of many massacres, pogroms, mass murders, indeed instances not far removed from genocide, that have been perpetrated without modern bureaucracy, the skills and technologies it commands, the scientific principles of its internal management. The Holocaust, however, was clearly unthinkable without such bureaucracy. The Holocaust was not an irrational outflow of the not-yet-fully-eradicated residues of pre-modern barbarity. It was a legitimate resident in the house of modernity; indeed, one who would not be at home in any other house.

This is not to suggest that the incidence of the Holocaust was *determined* by modern bureaucracy or the culture of instrumental rationality it epitomizes; much less still, that modern bureaucracy *must* result in Holocaust-style phenomena. I do suggest, however, that the rules of instrumental rationality are singularly incapable of preventing such phenomena; that there is nothing in those rules which disqualifies the Holocaust-style methods of 'social-engineering' as improper or, indeed, the actions they served as irrational. I suggest, further, that the bureaucratic culture which prompts us to view society as an object of administration, as a collection of so many 'problems' to be solved, as 'nature' to be 'controlled', 'mastered' and 'improved' or 'remade', as a legitimate target for 'social engineering', and in general a garden to be designed and kept in the planned shape by force (the gardening posture divides vegetation into 'cultured plants' to be taken care of, and weeds to be exterminated), was the very atmosphere in which the idea of the Holocaust could be conceived, slowly yet consistently developed, and brought to its conclusion. And I also suggest that it was the spirit of instrumental rationality, and its modern, bureaucratic form of institutionalization, which had made the Holocaust-style solutions not only possible, but eminently 'reasonable'—and increased the probability of their choice. This increase in probability is more than fortuitously related to the ability of modern bureaucracy to co-ordinate the action of great number of moral individuals in the pursuit of any, also immoral, ends....

Modern culture is a garden culture. It defines itself as the design for an ideal life and a perfect arrangement of human conditions. It constructs its own identity out of distrust of nature. In fact, it defines itself and nature, and the distinction between them, through its endemic distrust of spontaneity and its longing for a better, and necessarily artificial, order. Apart from the overall plan, the artificial *order* of the garden needs tools and raw materials. It also needs defence—against the unrelenting danger of what is, obviously, a disorder. The order, first conceived of as a design, determines what is a tool, what is a raw material, what is useless, what is irrelevant, what is harmful, what is a weed or a pest. It classifies all elements of the universe by their relation to itself. This relation is the only meaning it grants them and tolerates—and the only justification of the gardener's actions, as differentiated as the relations themselves. From the point of view of the design all actions are instrumental, while all the objects of action are either facilities or hindrances.

Modern genocide, like modern culture in general, is a gardener's job. It is just one of the many chores that people who treat society as a garden need to undertake. If garden design defines its weeds, there are weeds wherever there is a garden. And weeds are to be exterminated. Weeding out is a creative, not a destructive activity. It does not differ in kind from other activities which combine in the construction and sustenance of the perfect garden. All visions of society-as-garden define parts of the social habitat as human weeds. Like other weeds, they must be segregated, contained, prevented from spreading, removed and kept outside the society boundaries; if all these means prove insufficient, they must be killed.

Stalin's and Hitler's victims were not killed in order to capture and colonize the territory they occupied. Often they were killed in a dull, mechanical fashion with no human emotions—hatred included—to enliven it. They were killed because they did not fit, for one reason or another, the scheme of a perfect society. Their killing was not the work of destruction, but creation. They were eliminated, so that an objectively better human world—more efficient, more moral, more beautiful—could be established. A Communist world. Or a racially pure, Aryan world. In both cases, a harmonious world, conflict-free, docile in the hands of their rulers, orderly, controlled. People tainted with ineradicable blight of their past or origin could not be fitted into such an unblemished, healthy and shining world. Like weeds, their nature could not be changed. They could not be improved or re-educated. They had to be eliminated for reasons of genetic or ideational heredity—of a natural mechanism resilient and immune to cultural processing.

The two most notorious and extreme cases of modern genocide did not betray the spirit of modernity. They did not deviously depart from the main track of the civilizing process. They were the most consistent, uninhibited expressions of that spirit. They attempted to reach the most ambitious aims of the civilizing process most other processes stop short of, not necessarily for the lack of good will. They showed what the rationalizing, designing, controlling dreams and efforts of modern civilization are able to accomplish if not mitigated, curbed or counteracted.

These dreams and efforts have been with us for a long time. They spawned the vast and powerful arsenal of technology and managerial skills. They gave birth to institutions which serve the sole purpose of instrumentalizing human behavior to such an extent that any aim may be pursued with efficiency and vigor, with or without ideological dedication or moral approval on the part of the pursuers. They legitimize the rulers' monopoly on ends and the confinement of the ruled to the role of means. They define most actions as means, and means as subordination—to the ultimate end, to those who set it, to supreme will, to supra-individual knowledge.

Emphatically, this does not mean that we all live daily according to Auschwitz principles. From the fact that the Holocaust is modern, it does not follow that modernity is a Holocaust. The Holocaust is a by-product of the modern drive to a fully designed, fully controlled world, once the drive is getting out of control and running wild. Most of the time, modernity is prevented from doing so. Its ambitions clash with the pluralism of the human world; they stop short of their fulfillment for the lack of an absolute power absolute enough and a monopolistic agency monopolistic enough to be able to disregard, shrug off, or overwhelm all autonomous, and thus countervailing and mitigating, forces.

"Civilization Is Like a Jetliner" (1983)

T. Fulano

The night the Korean airliner crashed into the newspapers, I dreamed of a tornado. A tornado is a kind of spiral, which is the labyrinth and which is Death.

Death is very powerful right now. Instead of being a passage, Death has become a kind of equipment failure, a technical slaughterhouse. Human and technical failure become indistinguishable when the unquestioning robot and the drooling sadist merge. (I see the Soviet pilot being interviewed—he could be any Air Force gunslinger in any military machine—"I'd do it again—and even more and love every second of it." Of course he had the cooperation of the CIA and the U.S. military, who listened in, taping it all, without issuing any warnings to save lives. That, after all, is certainly not their business.)

So we inch closer to midnight. Death's festival. Reagan, on a California surfboard of lies and hypocritical self-righteousness, rides the crest triumphant, saying that the downing of the KAL 007 (how could it not be a spy plane with such a number!) represents "a major turning point" in world history, adding, "We can start preparing ourselves for what John F. Kennedy called a long twilight struggle." Another falsehood: crime flows into crime, from the extermination of the Indian "savages" to the wholesale massacres of Vietnamese "natives"—they've been fighting their twilight struggle for as long as anyone can remember, these evangelical maniacs, these scourges of the Great Darkness, these agents of Entropy.

But we must remember that the crash is representative, ultimately, of all air disasters, with its dash of militaristic insanity—in a sense, only a variant of the technological frenzy—thrown in for good measure. Civilization is like a jetliner, its East and West versions just the two wings, whose resistance holds the bulky, riveted monster aloft.

Civilization is like a jetliner, noisy, burning up enormous amounts of fuel. Every imaginable and unimaginable crime and pollution had to be committed in order to make it go. Whole species were rendered extinct, whole populations dispersed. Its shadow on the waters resembles an oil slick. Birds are sucked into its jets and vaporized. Every part—as Gus Grim once nervously remarked about space capsules before he was burned up in one—has been made by the lowest bidder.

Civilization is like a 747, the filtered air, the muzak oozing over the earphones, the phony sense of security, the chemical food, the plastic trays, all the passengers sitting passively in the orderly row of padded seats staring at Death on the movie screen. Civilization is like a jetliner, an idiot savant in the cockpit manipulating computerized controls built by sullen wage workers, and dependent for his directions on sleepy technicians high on amphetamines with their minds wandering to sports and sex.

Civilization is like a 747, filled beyond capacity with coerced volunteers—some in love with the velocity, most wavering at the abyss of terror and nausea, yet still seduced by advertising and propaganda. It is like a DC-10, so incredibly enclosed that you want to break through the tin can walls and escape, make your own way through the clouds, and leave this rattling, screaming fiend approaching its breaking point. The smallest error or technical failure leads to catastrophe, scattering your sad entrails like belated omens all over the runway; knocks you out of your shoes, breaks all your bones like eggshells.

(Of course, civilization is like many other things besides jets—always *things*—a chemical drainage ditch, a woodland knocked down to lengthen an airstrip or to build a slick new shopping mall where people can buy salad bowls made out of exotic tropical trees which will be extinct next week, or perhaps a graveyard for cars, or a suspension bridge which collapses because a single metal pin has shaken loose. Civilization is a hydra. There is a multitude of styles, colors, and sizes of Death to choose from.)

Civilization is like a Boeing jumbo jet because it transports people who have never experienced their humanity where they were to places where they shouldn't go. In fact, it mainly transports businessmen in suits with briefcases filled with charts, contracts, more mischief—businessmen who are identical everywhere and hence have no reason at all to be ferried about. And it goes faster and faster, turning more and more places into airports, the (un)natural habitat of businessmen.

It is an utter mystery how it gets off the ground. It rolls down the runway, the blinking lights along the ground like electronic scar tissue on the flesh of the earth, picks up speed and somehow grunts, raping the air, working its way up along the shimmering waves of heat and the trash blowing about like refugees fleeing the bombing of a city. Yes, it is exciting, a mystery, when life has been evacuated and the very stones have been murdered.

But civilization, like the jetliner, this freak phoenix incapable of rising from its ashes, also collapses across the earth like a million bursting wasps, flames spreading across the runway in tentacles of gasoline, Samsonite, and charred flesh. And always the absurd rubbish, Death's confetti, the fragments left to mock us lying along the weary trajectory of the dying bird—the doll's head, the shoes, eyeglasses, a beltbuckle.

Jetliners fall, civilizations fall, this civilization will fall. The gauges will be read wrong on some snowy day (perhaps they will fail). The wings, supposedly defrosted, will be too frozen to beat against the wind and the bird will sink like a millstone, first gratuitously skimming a bridge (because civilization is also like a bridge, from Paradise to Nowhere), a bridge laden, say, with commuters on their way to or from work, which is to say, to or from an airport, packed in their cars (wingless jetliners) like additional votive offerings to a ravenous Medusa.

Then it will dive into the icy waters of a river, the Potomac perhaps, or the River Jordan, or Lethe. And we will be inside, each one of us at our specially assigned porthole, going down for the last time, like dolls' heads encased in plexiglas.

"Industrial Society and
Its Future" (1995)

Unabomber (aka "FC")

177

ℜ eedless to say, the scenarios outlined above do not exhaust all the possibilities. They only indicate the kinds of outcomes that seem to us most likely. But we can envision no plausible scenarios that are any more palatable than the ones we've just described. It is overwhelmingly probable that if the industrial-technological system survives the next 40 to 100 years, it will by that time have developed certain general characteristics: Individuals (at least those of the "bourgeois" type, who are integrated into the system and make it run, and who therefore have all the power) will be more dependent than ever on large organizations; they will be more "socialized" than ever and their physical and mental qualities to a significant extent (possibly to a very great extent) will be those that are engineered into them rather than being the results of chance (or of God's will, or whatever); and whatever may be left of wild nature will be reduced to remnants preserved for scientific study and kept under the supervision and management of scientists (hence it will no longer be truly wild). In the long run (say a few centuries from now) it is likely that neither the human race nor any other important organisms will exist as we know them today, because once you start modifying organisms through genetic engineering there is no reason to stop at any particular point, so that the modifications will probably continue until man and other organisms have been utterly transformed.

178

Whatever else may be the case, it is certain that technology is creating for human beings a new physical and social environment radically different from the spectrum of environments to which natural selection has adapted the human race physically and psychologically. If man is not adjusted to this new environment by being artificially reengineered, then he will be adapted to it through a long and painful process of natural selection. The former is far more likely than the latter.

179

It would be better to dump the whole stinking system and take the consequences.

THE OLD WAY AND CIVILIZATION (1994)

TAMARACK SONG

The Old Way is the way of living common to the Native Peoples of The Earth, no matter what the era, culture or region. It is also the way of the plants, the animals, the Air, and the Water. It is the way all things natural were, are, and will be. This timeless Way is called "old" only by those who have abandoned it and now measure time in passing. Few Humans in this day know it or live it.

Civilization is the lifeway of Peoples who control and regiment the natural order. It is the current lifeway of most Humans, and of the animals, plants, and environments they have harnessed or domesticated to live it....

A Comparison

I recently met an Elder who gave me a most beautiful description of the Old Way in two words—*sharing* and *kindness*. Exploring the folds and reflections of those two words gives a full and lush view of that lifeway. The same day another Elder was speaking of the concerns she had for her grandchildren being exposed to the dominant culture. In elaborating on her fears, she gave a succinct, three-word definition of Civilization—*individualism*, (the accumulation of) *possessions*, and *commercialism*.

This dichotomy leaves little wonder that the Conquerors' first reaction to Native Peoples is often one of revulsion and sub-human classification. The Conquerors see them as crazy savages, fighting against all odds in a war they cannot win. The most spiritual often appear to be the most warlike. The intruders cannot grasp that the People are defending what they see as their clear right to follow Spirit. They are fighting for the very life and health of their Mother. They see it as better to die in Her defense than for them, and their generations to follow, to live a life of subjugation and encagement. Such a life would mean being forced not only to witness, but to be an active part in, the slow poisoning and dismemberment of the Sacred Mother-Source. In the end the People can find pride in losing, while the Civilized hordes can only find shame in winning.

Civilized People are still conquering Native People, though with the complexity of the contemporary world political-economic structure, perhaps not as conspicuously as in past centuries. With the consumption of every fast-food burger goes a chunk of South American Rainforest four times the area of my Lodge. (The Rainforest is one of the last holdouts of the People.) The purchase of every Japanese product pushes the Ainu—the indigenous (and Caucasian) Old Way Japanese People—closer to the sea on the last, northernmost island they inhabit.

Conquerors are prone to defining their morality quite narrowly, which helps justify their ways. For instance, they found it hard to reconcile the fact that the Hopi, whom they viewed as peaceful, agricultural, and very spiritual People, commonly had extramarital relations; while the Apache, whom they regarded as heartless plunderers, were morally conservative and very strict concerning mated fidelity. Even something as seemingly innocuous as dance was intolerable to the Conquerors; they could not accept it as being more than just social entertainment. (Native People, for whom dance is a central spiritual, psychological, and cultural expression, were equally surprised when they found out that Civilized dance was *just* social.)

Some primary distinctions between Civilized People and Native People: The Civilized change the world to suit themselves, while the Native adapt themselves to the world as it is; the Civilized are ever discontent with their present situation and dedicate their entire lives to changing it, while the Native are ever thankful for the beauty and bounty they find themselves immersed in; the Civilized dwell in the errors of the past and the hope of the future, while the Native bask in the fullness of the moment; the Civilized draw everything toward themselves while the Native become of everything about them; the Civilized grovel and beg as they contritely pray, while the Native pridefully sing in praise, thanksgiving, and wonderment; the Civilized have psychologists to help them adjust to their unreasonable lives, while the Native live in the harmony of their environs; the Civilized have religion, the Native live religion; the Civilized talk a lot, the Native listen and learn. The Civilized admire each other for *what* they are; the Native admire each other for *who* they are. The Civilized meet death lying in bed expending every effort to further extend life, while the Native greet death upright, if possible, with their Song of Passing on their lips as they greet the New Cycle.

Civilization is based on Human-made things that keep breaking down; the Old Way is based on natural things, which keep growing, renewing. Human-made things need regular input, while natural things keep giving. Civilized People become enslaved to their possessions, ever working to maintain them, while Native People are as free and unencumbered as the natural things that provide their needs.

Work as a concept is known only to Civilized People. It was born of the necessity to support the individualism and material opulence intrinsic to the lifeway. Where Natives avoid unnecessary duplication by sharing tools and other resources, Civilized People strive to individually possess whatever they use. They lead a catch-22 existence—they buy houses and cars so they can get jobs, then they have to keep their jobs so they can support their houses and cars. Their houses bulge with specialized rooms that are little used, while the lodges of the Natives are small and open, designed for multiple usage of space.

The material comparisons go on, but this will suffice to illustrate that Civilized People are working largely for things they don't use. They are committed to payments, taxes, insurance, maintenance, utility bills, and so on, no matter if or how much their material goods are used.

Their "labor-saving" devices actually save them little; the time saved is consumed by working elsewhere to pay for the tool, its fuel, maintenance, and the costs of storage. Some appliances, such as the washing machine, are not timesavers for an additional reason—their advent enabled more consumption. Now people have more clothes, and change and wash them more often, spending just as much time on laundry as before the machine.

Native People require but an average of two hours a day to provide their needs and desires, no matter whether the environment is lush tropic or desert. Their rich cultures, strong families, and lavish handiworks attest to their bountiful spare time. Their labor applies directly to their needs, as opposed to the more abstract Civilized concept of "going to work" to provide needs in a less direct way. Simply put, Natives transfer energy efficiently by direct involvement in what they need; whereas Civilized People, through a complex and non-personally involved process, expend much more time and energy to meet the same need. For instance, when the Native People desire fruit, they will simply go and pick it, whereas Civilized People will buy land, and go through the process of raising the fruit before picking it, or "go to work" to pay some-one else to raise (package, store and transport) it for them.

Those who have lived both Ways talk of the richer, more ful-filling life of the Native Way, with its direct involvement in the process of existence, as compared with the detached, indirect means of the Civilized Way. I first felt this difference when I was invited to share a meal with a Native family. The food had a life and a spirit that was given to it by their hands as they hunted, gar-dened, foraged, stored, prepared, and served it. This was reflected in the Blessing of the food, the way it was presented, eaten, and enjoyed, and in the way it was valued and respected, without a bite being wasted. What a blessed experience when compared to my hollow store-bought meals!

There is little sacred in Civilized societies. They are systems-oriented; they look to structure for answers, not knowing of the ways of Elders and the Talking Circle and the Inner Voice. The once-sacred becomes lowered to the Civilized society's secular norm. Drugs, alcohol, and sex become objects of pleasure, where in the few Native societies where drugs or alcohol are used, they are used sporadically, and as part of sacred rituals.

Civilized People are ego-sensitive; self-recountings of their adventures and successes often come across as self-aggrandizing and ego-threatening to the listener. In cultures where the Warrior and the Healer and the Seeker still exist, stories of their Journeys and triumphs are regularly told and eagerly awaited. Beyond entertain-ment, these recountings serve as teachers and examples to inspire and emulate. Perhaps because Native People have more opportuni-ty for self-fulfillment than their Civilized Counterparts, they are less threatened and more inspired by the success of others.

My impression is that the unspoken Civilized objective is to fashion an Earth (and beyond?) that is under total Human control. What Native People see as their natural realm, Civilized People see as uncontrolled, wild. Their neighbors are no longer the animals and plant People, but other Humans. So the natural realm is truly wild to them, and their isolation from it isolates them from its care, and from its wisdoms. For an example, with many non-Human People, staring into another's eyes is a sign of assertiveness and dominance, or of aggression. It is also a giveaway to the stalked and a preoccupation that puts one out of contact with the Greater Cir-cle. For these reasons, Natives consider it foolish and disrespectful to stare into the eyes of another, particularly an Elder. No longer knowing the animals to gather these lessons, Civilized People suffer interactions plagued with the friction of their eyes and the imbalance of their perception.

Civilized Peoples' care for their Source of their goods is not sensitive to Her needs because they do not know Her needs. For example, when logging for their lumber and paper, they don't know to let some of the big, hollow trees stand, so one-quarter of the varieties of our bird kin are left homeless.

One reason for the "success" of the Civilized Way is its willingness to adopt the ways of other cultures that work to its advantage. This approach has created functional cultures but without the Ancestral roots and spiritual bases of the cultures from which they borrow. For example, they have borrowed practices from the people of India, such as they call Yoga and Transcendental Meditation. They are fragments of a Hindu People's life approach, surface techniques which are a reflection of the underlying philosophy. Only the exercise is desired; its spirit is left behind. This allows for Civilization's penchant to commercially exploit other cultures. So we see these borrowed practices being promoted with such lines as "reduce stress, increase productivity, lose weight, be a better yuppie or salesman by practicing..." The Civilized Way could benefit greatly from a deeper look at and understanding of the ways of other cultures, but instead it is content skimming the grease off the top and using it to lubricate the worn-out mechanism of its lifeway.

The Civilized Way can be characterized by such contemporary clichés as "the me generation," "self-development," and "I do my thing, you do yours." The most powerful contemporary response that I've heard is Albert Schweitzer's, which echoes Old Way wisdom, "Life outside a person is an extension of the life within him. This compels him to be part of it and accept responsibility for all creatures great and small. Life becomes harder when we live for others, but it also becomes richer and happier."

Where They Diverged

If we are all the same People, where did our Paths diverge and some of us turn from The Mother to see if we could do better? Perhaps the answer lies in the way we look at a seed. Agriculture is the basis of Civilization; with it came permanent settlements and the concept of land ownership. The Earth became "property"—a despiritualized, inanimate commodity. Now Civilization had a foundation upon which to lay its cornerstones—the concentration of wealth and power, predatory trade and warfare, and the enslavement of humans, animals, plants, Water, and minerals.

The Old Way, based on foraging economies dependent upon a respectful relationship with Earth, can give no root or nourishment to the above-mentioned Civilized traits. Nor can it support cultural, economic, and political stratification. Instead, its small interactive groups, which share in spirit, strife, and pleasure, encourage a more personally involved, less bounded lifeway.

Native village soil-tillers became the transitional step between the Old Way and Civilization. It is here that we first see powerful leaders, class systems, and wealthy individuals. It is also here where interest, rent, currency, and animal and human sacrifice make their entrance, as they are largely absent from the lifeways of foraging Peoples.

Women/Wilderness (1986)

Ursula K. LeGuin

What Freud mistook for her lack of civilisation is woman's lack of loyalty to civilisation.

—Lillian Smith

Civilized Man says: I am Self, I am Master, all the rest is Other—outside, below, underneath, subservient. I own, I use, I explore, I exploit, I control. What I do is what matters. What I want is what matter is for. I am that I am, and the rest is women and the wilderness, to be used as I see fit.

To this, Civilized Woman, in 1978, in the voice of Susan Griffin, replies as follows:

> We say there is no way to see his dying as separate from her living, or what he had done to her, or what part of her he had used. We say if you change the course of this river you change the shape of the whole place. And we say that what she did then could not be separated from what she held sacred in herself, what she had felt when he did that to her, what we hold sacred to ourselves, what we feel we could not go on without, and we say if this river leaves this place, nothing will grow and the mountain will crumble away, and we say that what he did to her could not be separated from the way that he looked at her, and what he felt was right to do to her, and what they do to us, we say, shapes how they see us. That once the trees are cut down, the water will wash the mountain away and the river be heavy with mood, and there will be a flood. And we say that what he did to her he did to all of us. And that one act cannot be separated from another. And had he seen more clearly, we say, he might have predicted his own death. How if the trees grew on the hillside there would be no flood. And you cannot divert the river. We say look how the water

flows from this place and returns as rainfall, everything returns, we say, and one thing follows another, there are limits, we say, on what can be done and everything moves. We are all part of this motion, we say, and the way of the river is sacred, and this grove of trees is sacred, and we ourselves, we tell you, are sacred.

—Susan Griffin, *Woman and Nature*

What is happening here is that the wilderness is answering. This has never happened before. We who live at this time are hearing news that has never been heard before. A new thing is happening.

Daughters, the women are speaking
They arrive
over the wise distances
on perfect feet.

The women are speaking: so says Linda Hogan of the Chickasaw people. The women are speaking. Those who were identified as having nothing to say, as sweet silence or monkey chatterers, those who were identified with Nature, which listens, as against Man, who speaks—those people are speaking. They speak for themselves and for the other people, the others who have been silent, or silenced, or unheard, the animals, the trees, the rivers, the rocks. And what they say is: We are sacred.

Listen: they do not say "Nature is sacred." Because they distrust that word, Nature. Nature as not including humanity, Nature as what is not human, that Nature is a construct made by Man, not a real thing; just as most of what Man says and knows about women is mere myth and construct. Where I live as a woman is to men a wilderness. But to me it is home.

The anthropologists Shirley and Edwin Ardener, talking about an African village culture, made a useful and interesting mental shape. They laid down two circles largely but not completely overlapping, so that the center of the figure is the tall oval of overlap, and on each side of it are facing crescents of non-overlap. One of the two circles is the dominant element of the culture, that is, Men. The other is the Muted element of the culture, that is, Women. As Elaine Showalter explains the figure, "All male consciousness is within the dominant circle, accessible to or structured by language." Both the crescent that belongs to men only, and the crescent that belongs to women only, outside the shared, central, civilized area of overlap, may be called "the wilderness." The men's wilderness is real, it is where men can go hunting and exploring and having all-male adventures, away from the village, the shared center, and it is accessible to and structured by language. "In terms of cultural anthropology, women know what the male crescent is like, even if they have never seen it, because it becomes the subject of legend ... But men do not know what is in the wild," that is, the no-man's-land, the crescent that belongs to the muted group, the silent group, the group within the culture that *is not spoken*, whose experience is not considered to be part of human experience, that is, the women.

Men live their whole lives within the Dominant area. When they go off hunting bears, they come back with bear stories, and these are listened to by all, they become the history or the mythology of that culture. So the men's "wilderness" becomes "Nature," considerd as the property of "Man."

But the experience of women as women, their experience unshared with men, that experience is the wilderness or the wildness that is utterly other—that is, in fact, to Man, unnatural. That is what civilization has left out, what culture excludes, what the Dominants call animal, bestial, primitive, undeveloped, unauthentic ... what has not been spoken, and when spoken, has not been heard ... what we are just beginning to find words for, our words, not their words: the experience of women. For dominance-identified men and women both, that is true wildness. Their fear of it is ancient, profound, and violent. The misogyny that shapes every aspect of our civilization is the institutionalized form of male fear and hatred of what they have denied, and therefore cannot know, cannot share: that wild country, the being of women.

All we can do is try to speak it, try to say it, try to save it. Look, we say, this land is where your mother lived and where your daughter will live. This is your sister's country. You lived there as a child, boy or girl, you lived there—have you forgotten? All children are wild. You lived in the wild country. Why are you afraid of it?

SECTION IV

———

THE PATHOLOGY OF
CIVILIZATION

the flatiron builving

Set down there not knowing it was Seattle, I could not have told where I was. Everywhere frantic growth, a carcinomatous growth. Bulldozers rolled up the green forests and heaped the resulting trash for burning. The torn white lumber from concrete forms was piled beside gray walls. I wonder why progress looks so much like destruction.

—John Steinbeck (1962)

This section cannot be cleanly separated from the preceding one. The distinction, such as it is, is one of emphasis. From pictures that help to reveal basic qualities of civilization, we shift in the direction of focusing on civilization's dynamics and what they portend, now and in the future. Fully established, mature civilization is what has to be grasped, in its malignant and metastasizing trajectory.

There is little doubt what is in store: a steadily bleaker and more debased reality, with civilization's ideological defenses eroding to naught. Of course, there have always been some who could see through the massive fraud. Consider William Morris, writing in 1885, a banner year for ascendant industrial capitalism:

I have no more faith than a grain of mustard seed in the future of "civilisation," which I know now is doomed to destruction, and probably before very long; what a joy it is to think of! and how often it consoles me to think of barbarism once more flooding the world, and real feelings and passions, however rudimentary, taking the place of our wretched hypocrisies.... I used really to despair because I thought what the idiots of our day call progress would go on perfecting itself: happily I know that all will have a sudden check.

He was overly optimistic about the imminence of civilization's downfall, but the ranks of its critics, if this collection is any gauge at all, have certainly swelled since Morris registered his judgment.

Even some of the high priests of civilization have abandoned earlier enthusiasm or faith in its latest triumphs. In the contemporary era of high-tech mania and belief in the transcendent contributions of instantaneous computerized interaction, critics are beginning to multiply. By the mid-1970s even Marshall McLuhan came to some very uncelebratory conclusions. For example:

> Electronic media reduce personal identity to vestigial levels that, in turn, diminish moral feeling to practically nothing.

Other critics have recognized that postmodernism, far and away the reigning cultural zeitgeist, plays an essential and duplicitous role in the defense of civilization. Qualities like cynicism, relativism, and superficiality are part of this, but the postmodern gloss on society goes even further in its efforts to deflect opposition to civilized social existence. Frederic Jameson captures this aptly when he asks

> How it is possible for the most standardized and uniform social reality in history, by the merest ideological flick of the thumbnail, the most imperceptible of displacements, to reemerge as the rich oil-smear sheen of absolute diversity and the unimaginable and unclassifiable forms of human freedom.

Postmodernism seems to go beyond mere denial, to actually affirm our ghastly present. Aversion to analysis, a key postmodern trait, can and does obscure that which needs to be seen for what it is, and confronted.

Conventional Lies, or
Our Civilization (1895)

Max Nordau

This universal mental restlessness and uneasiness exerts a powerful and many-sided influence upon individual life. A dread of examining and comprehending the actualities of life prevails to a frightfully alarming extent, and manifests itself in a thousand ways. The means of sensation and perception are eagerly counterfeited by altering the nervous system by the use of stimulating or narcotic poisons of all kinds, manifesting thereby an instinctive aversion to the realities of appearances and circumstances. It is true that we are only capable of perceiving the changes in our own organism, not those going on around us. But the changes within us are caused, most probably, by objects outside of us; our senses give us a picture of those objects, whose reliability is surely more to be depended upon, when only warped by the imperfections in our normal selves, than when to these unavoidable sources of error is added a conscious disturbance in the functions of the nervous system caused by the use of various poisons. Only when our perceptions of things around us awake in us a feeling of positive discomfort, do we realize the necessity of warding off these unpleasant sensations, or of modifying them, until they become more agreeable. This is the cause of the constant increase in the consumption of alcohol and tobacco, shown by statistics, and of the rapidity with which the custom of taking opium and morphine is spreading. It is also the reason why the cultivated classes seize upon every new narcotic or stimulant which science discovers for them, so that we have not only drunkards and opium eaters among us, but confirmed chloral, chloroform and ether drinkers. Society as a whole repeats the action of the individual, who tries to "drown his sorrows in the flowing bowl." It seeks oblivion of the present, and grasps at anything that will provide it with the necessary illusions by which it can escape from real life.

Hand in hand with this instinctive self-deception and attempt at temporary oblivion of the actual world, goes the final plunge into eternal oblivion: statistics prove that the number of suicides is increasing in the highly civilized countries, in direct proportion to the increase in the use of alcohol and narcotics. A dull sensation of irritation, sometimes self-conscious, but more often only recognized as a vague, irresistible discontent, keeps the aspiring in a state of gloomy restlessness, so that the struggle for existence assumes brutal and desperate phases, never known before. This struggle is no longer a conflict between polite antagonists who salute each other with courtesy before they open fire, like the English and French before the battle of Fontenoy, but it is a pell-mell, hand-to-hand fight of rough cut-throats, drunk with whisky and blood, who fall upon each other with brute ferocity, neither giving nor expecting mercy. We lament the disappearance of characters. What is a character? It is an individuality which shapes its career according to certain simple, fundamental moral principles which it has recognized as good, and accepted as guides. Scepticism develops no such characters, because it has excluded faith in fundamental principles. When the north star ceases to shine, and the electric pole vanishes, the compass is of no further use the stationary point is gone—to which it was always turning. Scepticism, also a fashionable ailment, is in reality but another phase of the universal discontent with the present.

For it is only by becoming convinced that the world is out of sorts generally, and that everything is wrong, insufficient, and contemptible, that we arrive at the conclusion that all is vanity, and nothing worth an effort, or a struggle between duty and inclination. Economy, literature, and art, philosophy, politics, and all phases of social and individual life, show a certain fundamental trait, common to all—a deep dissatisfaction with the world as it exists at present. From each one of these multitudinous manifestations of human intelligence arises a bitter cry, the same in all cases, an appeal for a radical change.

The Final Empire:
The Collapse of Civilization and
The Seed of the Future (1993)

William H. Koetke

Our generation is on the verge of the most profound catastrophe the human species has ever faced. Death threats to the living earth are coming from all sides. Water, sunlight, air and soil are all threatened. When Eskimos of the far north begin to experience leukemia from atomic radiation and Eskimo mothers' milk contains crisis levels of PCBs, we must recognize that every organism on the planet is threatened.

Compounding this crisis is the fact that the prime force in this affair, the civilized humans, are unable to completely understand the problem. The problem is beneath the threshold of consciousness because humans within civilization (civilization comes from the Latin, *civis*, referring to those who live in cities, towns and villages) no longer have relationship with the living earth. Civilized people's lives are focused within the social system itself. They do not perceive the eroding soils and the vanishing forests. These matters do not have the immediate interest of paychecks. The impulse of civilization in crisis is to do what it has been doing, but do it more energetically in order to extricate itself. If soaring population and starvation threaten, often the impulse is to put more pressure on the agricultural soils and cut the forests faster.

We face planetary disaster. The destruction of the planetary life system has been ongoing for thousands of years and is now approaching the final apocalypse which some of us will see in our own lifetimes. Far from being a difficult and complex situation it is actually very simple, if one can understand and accept a few simple and fundamental propositions.

The planetary disaster is traced to one simple fact. *Civilization is out of balance with the flow of planetary energy.* The consensus assumption of civilization is that an exponentially expanding human population with exponentially expanding consumption of material resources can continue, based on dwindling resources and a dying ecosystem. This is simply absurd. Nonetheless, civilization continues on with no memory of its history and no vision of its future.

Possibly the most important source of life on this planet is the thin film of topsoil. The life of the planet is essentially a closed, balanced system with the elements of sun, water, soil and air as the basic elements. These elements work in concert to produce life and they function according to patterns that are based in the laws of physics, which we refer to as Natural Law.

The soil depth and its richness is a basic standard of health of the living planet. As a general statement we may say that when soil is lost, imbalance and injury to the planet's life occurs. In the geologic time-span of the planet's life, this is a swift progression toward death. Even if only one per cent of the soil is lost per thousand years, eventually the planet dies. If one per cent is gained, then the living wealth, the richness, of the planet increases. The central fact must be held in mind of how slowly soil builds up. Soil scientists estimate that 300 to 1000 years are required for the buildup of each inch of topsoil.

The nourishment of the soil depends upon the photosynthetic production of the vegetative cover that it carries. There are wide differences in the Net Photosynthetic Production of many possible vegetative covers. As a rule it is the climax ecosystem of any particular region of the earth that is the most productive in translating the energy of the sun into the growth of plants and in turn into organic debris which revitalizes the soil.

A climax ecosystem is the equilibrium state of the "flesh" of the earth. After a severe forest fire, or to recover from the injury of clearcut logging, the forest organism slowly heals the wound by inhabiting the area with a succession of plant communities. Each succeeding community prepares the area for the next community. In general terms, an evergreen forest wound will be covered by tough small plants, popularly called "weeds" and the grasses which hold down the topsoil and prepare the way for other grasses and woody shrubs to grow up on the wound. ("Weeds" are the "first aid crew" on open ground.) As a general rule, the "first aid crew"—the first community of plants to get in and cover the bare soil and hold it down—is the more simple plant community with the smallest number of species of plants, animals, insects, micro-organisms and so forth. As the succession proceeds, the diversity, the number of species, increases as does the NPP, until the climax system is reached again, and equilibrium is established. *The system drives toward complexity of form, maximum ability to translate incoming energy (NPP) and diversity of energy pathways (food chains and other services that plants and animals perform for one another).* The plants will hold the soil so that it may be built back up. They will shade the soil to prevent its oxidation (the heating and drying of soil promotes chemical changes that cause sterility) and conserve moisture. Each plant takes up different combinations of nutrients from the soil so that specific succession communities prepare specific soil nutrients for specific plant communities that will succeed them. Following the preparation of the site by these plants, larger plants, alders and other broadleaf trees will come in and their lives and deaths will further prepare the micro-climate and soil for the evergreens. These trees function as "nurse" trees for the final climax community, which will be conifers. Seedling Douglas Fir, for example, cannot grow in sunlight and must have shade provided by these forerunner communities.

The ecosystems of this earth receive injury from tornado, fire, or other events and then cycle back to the balanced state, the climax system. This is similar to the wound on a human arm that first bleeds, scabs over and then begins to build new replacement skin to reach its equilibrium state. *The climax system then is a basic standard of health of the living earth, its dynamic equilibrium state. The climax system is the system that produces the greatest photosynthetic production.* Anything that detracts from this detracts from the health of the ecosystem.

Climax ecosystems are the most productive because they are the most diverse. Each organism feeds back some portion of energy to producers of energy that supports it (as well as providing energy to other pathways) and as these support systems grow, the mass and variety of green plants and animals increases, taking advantage of every possible niche. What might be looked at as a whole, unitary organ of the planet's living body, a forest or grassland, experiences increased health because of its diversity within.

On a large scale, the bioregions and continental soils substantially support sea life by the wash-off (natural and unnatural) of organic fertility into aquatic and ocean environments. This is a further service that these whole ecosystems perform for other whole ecosystems.

A few basic principles of the earth's life in the cosmos have now been established. Balance is cosmic law. The earth revolves around the sun in a finely tuned balance. The heat budget of the planet is a finely tuned balance. If the incoming heat declined, we would freeze or if the planet did not dissipate heat properly we would burn up. The climax ecosystem maintains a balance and stability century after century as the diverse flows of energies constantly move and cycle within it. In the same manner the human body maintains balance (homeostasis) while motion of blood, digestion and cell creation, flow within it.

The life of the earth is fundamentally predicated upon the soil. If there is no soil, there is no life as we know it. (Some micro-organisms and some other forms might still exist.) *The soil is maintained by its vegetative cover and in optimal, balanced health, this cover is the natural climax ecosystem.*

If one can accept these few simple principles then we have established a basis of communication upon which we may proceed. Anyone who cannot accept these principles must demonstrate that the world works in some other way. This must be done quickly because the life of the planet earth hangs in the balance.

We speak to our basic condition of life on earth. We have heard of many roads to salvation. We have heard that economic development will save us, solar heating will save us, technology, the return of Jesus Christ who will restore the heaven and the earth, the promulgation of land reform, the recycling of materials, the establishment of capitalism, communism, socialism, fascism, Muslimism, vegetarianism, trilateralism, and even the birth of new Aquarian Age, we have been told, will save us. *But the principle of soil says that if the humans cannot maintain the soil of the planet, they cannot live here.* In 1988, the annual soil loss due to erosion was 25 billion tons and rising rapidly. Erosion means that soil moves off the land. An equally serious injury is that the soil's fertility is exhausted in place. Soil exhaustion is happening in almost all places where civilization has spread. This is a literal killing of the planet by exhausting its fund of organic fertility that supports other biological life. Fact: since civilization invaded the Great Plains of North America one-half of the topsoil of that area has disappeared.

The Record of Empire

The eight-thousand-year record of crimes against nature committed by civilization include assaults on the topsoils of all continents.

Forests, the greatest generators of topsoil, covered roughly one-third of the earth prior to civilization. By 1975 the forest cover was one-fourth, and by 1980 the forest had shrunk to one-fifth and the rapidity of forest elimination continues to increase. If the present trends continue without interruption 80 percent of the vegetation of the planet will be gone by 2040.

The simple fact is that civilization cannot maintain the soil. Eight thousand years of its history demonstrate this. Civilization is murdering the earth. The topsoil is the energy bank that has been laboriously accumulated over millennia. Much of it is gone and the remainder is going rapidly.

When civilized "development" of land occurs the climax system is stripped, vegetation is greatly simplified or cleared completely, and the net photosynthetic production plummets. In the tropics, when pasture land is created by clearing forest, two-thirds of the original net photosynthetic production is eliminated. In the mid-latitudes one-half the net photosynthetic production is lost when cropland is created from previously forested land. The next step is that humans take much of even that impaired production off the land in the form of agricultural products so that not even the full amount of that impaired production returns to feed the soil.

This points out a simple principle: *Human society must have as its central value, a responsibility to maintain the soil.* If we can create culture that can maintain the soil then there is the possibility of human culture regaining balance with the life of the earth.

The central problem is that civilization is out of balance with the life of the earth.

The solution to that problem is for human society to regain balance with the earth.

We are now back to everyone's personal answer concerning how to respond to the planetary crisis. Most proposals for salvation have little to do with maintaining the soil. All of these seek to alleviate the situation without making any uncomfortable change in the core values or structure of existing society. They only try to "fix" the symptoms. If we had a society whose core values were to preserve and aid the earth, then all of the other values of society would flow consistently from that.

In many important ways civilization functions in an addictive fashion. The culture of civilization functions so that it is self-destructive, suicidal; as if it were a person addicted to alcohol, white sugar, drugs or tobacco. The addict denies that there is a problem. The addict engages in the denial of reality. Civilization is addicted in the same way.

The civilized people believe they have an obligation to bring primitive and underdeveloped people up to their level. Civilization, which is about to self-destruct, thinks of itself as the superior culture that has answers for all the world's people.

An addict, truly, is a person who is emotionally dependent on things: television, substances, personality routines, other people, mental ideologies, total immersion in some cause or work. If the object of dependency is removed, addicts will experience insecurity, discomfort, distress, the symptoms of withdrawal.

Civilization is a cultural/mental view that believes security is based in instruments of coercion. The size of this delusion is such that the combined military expenditures of all the world's governments in 1987 were so large that all of the social programs of the United Nations could be financed for 300 years by this expenditure.

Looking back at the simple principle which says that humans cannot live on this planet unless they can maintain the topsoil, demonstrates the delusion. The civilized denial of the imperative of maintaining topsoil, demonstrates the delusion. The delusion of military power does not lead to security, it leads to death. The civilized denial of the imperative of maintaining topsoil, and the addictive grasping to the delusion that security can be provided by weapons of death, is akin to the hallucination of an alcoholic suffering delirium tremens!

The first step in the recovery of any addict is the recognition that what they have believed is a delusion. The alcoholic must come to see that "just one more drink" is not the answer, the workaholic must come to see that "just a little more effort" will not provide feelings of self-worth and a rounded life. The bulimic must come to see that "just one more plate of food" will not provide emotional wholeness. *Civilization must come to see that its picture of reality is leading it to suicide....*

Here we have the whole of it. *The problem is imbalance and the solution is to regain balance.* Here we have the simple principle: if human actions help to regain balance as judged by the condition of the soil, then we are on the path of healing the earth. If the theory, plan, project, or whatever, cannot be justified by this standard, then we are back in the delusional system.

All of us are addicts. We of civilization have lost our way. We are now functioning in a world of confusion and chaos. We must recognize that the delusional system of civilization, the mass institutions and our personal lives, function on a self-destructive basis. We live in a culture that is bleeding the earth to death, and we have been making long-range personal plans and developing careers within it. We strive toward something that is not to be.

We must try to wake up and regain a vision of reality. We must begin taking responsibility for our lives and for the soil. This is a tall order. This will require study and forethought. Humans have never dealt with anything like this before. This generation is presented with a challenge that in its dimensions is cosmic. A cosmic question: will tens of millions of years of the proliferation of life on earth die back to the microbes? This challenge presents us with the possibility of supreme tragedy or the supreme success.

Creating a utopian paradise, a new Garden of Eden is our only hope. Nothing less will extricate us. We must create the positive, cooperative culture dedicated to life restoration and then accomplish that in perpetuity, or we as a species cannot be on earth.

The Collapse of Complex Societies (1988)

Joseph A. Tainter

Understanding collapse: the marginal productivity of sociopolitical change

Not only is energy flow required to maintain a sociopolitical system, but the amount of energy must be sufficient for the complexity of that system. Leslie White observed a number of years ago that cultural evolution was intricately linked to the quantities of energy harvested by a human population (1949: 363–93). The amounts of energy required per capital to maintain the simplest human institutions are incredibly small compared with those needed by the most complex. White once estimated that a cultural system activated primarily by human energy could generate only about 1/20 horsepower per capita per year (1949:369. 1959:41–2). This contrasts sharply with the hundreds to thousands of horsepower at the command of the members of industrial societies. Cultural complexity varies accordingly. Julian Steward pointed out the quantitative difference between the 3,000 to 6,000 cultural elements early anthropologists documented for native populations of western North America, and the more than 500,000 artifact types that U.S. military forces landed at Casa Blanca in World War II (1955:81).

More complex societies are more costly to maintain than simpler ones, requiring greater support levels per capita. As societies increase in complexity, more networks are created among individuals, more hierarchical controls are created to regulate these networks, more information is processed, there is more centralization of information flow, there is increasing need to support specialists not directly involved in resource production, and the like....

Complex societies, such as states, are not a discrete stage in cultural evolution. Each society represents a point along a continuum from least to most complex. Complex forms of human organization have emerged comparatively recently, and are an anomaly of history. Complexity and stratification are oddities when viewed from the full perspective of our history, and where present, must be constantly reinforced. Leaders, parties and governments need constantly to establish and maintain legitimacy. This effort must have a genuine material basis, which means that some level of responsiveness to a support population is necessary. Maintenance of legitimacy or investment in coercion require constant mobilization of resources. This is an unrelenting cost that any complex society must bear....

There are major differences between the current and the ancient worlds that have important implications for collapse. One of these is that the world today is full. That is to say, it is filled by complex societies; these occupy every sector of the globe, except the most desolate. This is a new factor in human history. Complex societies as a whole are a recent and unusual aspect of human life. The current situation, where all societies are so oddly constituted, is unique. It was shown earlier in this chapter that ancient collapses occurred, and could only occur, in a power vacuum, where a complex society (or cluster of peer polities) was surrounded by less complex neighborhoods. There are no power vacuums left today. Every nation is linked to, and influenced by, the major powers, and most are strongly linked with one power bloc or the other. Combine this with instant global travel, and as Paul Valery noted, '... *nothing can ever happen again without the whole world's taking a hand*' (1962: 115 [emphasis in original]).

Collapse today is neither an option nor an immediate threat. Any nation vulnerable to collapse will have to pursue one of three options: (1) absorption by a neighbor or some larger state; (2) economic support by a dominant power, or by an international financing agency; or (3) payment by the support population of whatever costs are needed to continue complexity, however detrimental the marginal return. A nation today can no longer unilaterally collapse, for if any national government disintegrates its population and territory will be absorbed by some other.

Although this is a recent development, it has analogies in past collapses, and these analogies give insight into current conditions. Past collapses, as discussed, occurred among two kinds of international political situations; isolated, dominant states, and clusters of peer polities. The isolated, dominant state went out with the advent of global travel and communication, and what remains now are competitive peer polities. Even if today there are only two major peers, with allies grouped into opposing blocs, the dynamics of the competitive relations are the same. Peer polities, such as post-Roman Europe, ancient Greece and Italy, Warring States China, and the Mayan cities, are characterized by competitive relations, jockeying for position, alliance formation and dissolution, territorial expansion and retrenchment, and continual investment in military advantage. An upward spiral of competitive investment develops, as each polity continually seeks to outmaneuver its peer(s). None can dare withdraw from this spiral, without unrealistic diplomatic guarantees, for such would be only invitation to domination by another. In this sense, although industrial society (especially the United States) is sometimes likened in popular thought to ancient Rome, a closer analogy would be with the Mycenaeans or the Maya.

Peer polity systems tend to evolve toward greater complexity in a lockstep fashion as, driven by competition, each partner imitates new organizational, technological, and military features developed by its competitor(s). The marginal return on such developments declines, as each new military breakthrough is met by some countermeasure, and so brings no increased advantage or security on a lasting basis. A society trapped in a competitive peer polity system must invest more and more for no increased return, and is thereby economically weakened. And yet the option of withdrawal or collapse does not exist. So it is that collapse (from declining marginal returns) is not in the *immediate* future for any contemporary nation. This is not, however, due so much to anything we have accomplished as it is to the competitive spiral in which we have allowed ourselves to become trapped....

In ancient societies the solution to declining marginal returns was to capture a new energy subsidy. In economic systems activated largely by agriculture, livestock, and human labor (and ultimately by solar energy), this was accomplished by territorial expansion. Ancient Rome and the Ch'in of Warring States China adopted this course, as have countless other empire-builders. In an economy that today is activated by stored energy reserves, and especially in a world that is full, this course is not feasible (nor was it ever permanently successful). The capital and technology available must be directed instead toward some new and more abundant source of energy. Technological innovation and increasing productivity can forestall declining marginal returns only so long. A new energy subsidy will at some point be essential.

It is difficult to know whether world industrial society has yet reached the point where the marginal return for its overall pattern of investment has begun to decline. The great sociologist Pitirim Sorokin believed that Western economies had entered such a phase in the early twentieth century (1957: 530). Xenophon Zolotas, in contrast, predicts that this point will be reached soon after the year 2000 (1981: 102–3). Even if the point of diminishing returns to our present form of industrialism has not yet been reached, that point will inevitably arrive. Recent history seems to indicate that we have at least reached declining returns for our reliance on fossil fuels, and possibly for some raw materials. A new energy subsidy is necessary if a declining standard of living and a future global collapse are to be averted. A more abundant form of energy might not reverse the declining marginal return on investment in complexity, but it would make it more possible to finance that investment.

In a sense the lack of a power vacuum, and the resulting competitive spiral, have given the world a respite from what otherwise might have been an earlier confrontation with collapse. Here indeed is a paradox: a disastrous condition that all decry may force us to tolerate a situation of declining marginal returns long enough to achieve a temporary solution to it. This reprieve must be used rationally to seek for and develop the new energy source(s) that will be necessary to maintain economic well-being. This research and development must be an item of the highest priority, even if, as predicted, this requires reallocation of resources from other economic sectors. Adequate funding of this effort should be included in the budget of every industrialized nation (and the results shared by all). I will not enter the political foray by suggesting whether this be funded privately or publicly, only that funded it must be.

There are then notes of optimism and pessimism in the current situation. We are in a curious position where competitive interactions force a level of investment, and a declining marginal return, that might ultimately lead to collapse except that the competitor who collapses first will simply be dominated or absorbed by the survivor. A respite from the threat of collapse might be granted thereby, although we may find that we will not like to bear its costs. If collapse is not in the immediate future, that is not to say that the industrial standard of living is also reprieved. As marginal returns decline (a process ongoing even now), up to the point where a new energy subsidy is in place, the standard of living is also reprieved. As marginal returns decline (a process ongoing even now), up to the point where a new energy subsidy is in place, the standard of living that industrial societies have enjoyed will not grow so rapidly, and for some groups and nations may remain static or decline. The political conflicts that this will cause, coupled with the increasingly easy availability of nuclear weapons, will create a dangerous world situation in the foreseeable future.

To a degree there is nothing new or radical in these remarks. Many others have voiced similar observations on the current scene, in greater detail and with greater eloquence. What has been accomplished here is to place contemporary societies in a historical perspective, and to apply a global principle that links the past to the present and the future. However much we like to think of ourselves as something special in world history, in fact industrial societies are subject to the same principles that caused earlier societies to collapse.

If civilization collapses again, it will be from failure to take advantage of the current reprieve, a reprieve paradoxically both detrimental and essential to our anticipated future.

Where the Wasteland Ends: Politics and Transcendence in Postindustrial Society (1972)

Theodore Roszak

The Great Divide

If it seems cranky to lament the expanding artificiality of our environment, the fact underlying that lament is indisputable, and it would be blindness to set its significance at less than being the greatest and most rapid cultural transition in the entire history of mankind. This is the historical great divide—in one sense, quite literally. In little more than a century, millions of human beings in Europe and America—and their number grows daily throughout the world—have undertaken to divide themselves off more completely and irremediably from the natural continuum and from all that it has to teach us of our relationship to the non-human, than ever before in the human past.

It is all too easy to obscure this pre-eminent truth by conjuring up a picture of the remaining wide-open spaces—the mountain vastnesses and desert solitudes, the faraway islands and jungle thickets—and then to conclude, consolingly, that the cities will never encroach upon these remote corners of the earth. But that is wishful thinking already belied by fact and supported only by a misconception about the way in which urban-industrialism asserts its dominance. True enough, urban sprawl may never swallow these outlying areas into its concrete and steel maw. But that is not the only way the supercity propagates its power.

Before industrialism, most cities stood apart as modest workshops or markets whose ethos was bounded by their own walls. They were an option in the world, one way of life among many possibilities. The supercity, however—or rather the artificial environment taken as a whole—stretches out tentacles of influence that reach thousands of miles beyond its already sprawling perimeters. It sucks every hinterland and wilderness into its technological metabolism. It forces rural populations off the land and replaces them with vast agra-industrial combines. Its investments and technicians muscle their way into the back of every beyond, bringing the roar of the bulldozer and oil derrick into the most uncharted quarters. It runs its conduits of transport and communication, its lines of supply and distribution through the wildest landscapes. It flushes its wastes into every river, lake and ocean, or trucks them away into desert areas. The world becomes its garbage can—including the capacious vault of the atmosphere itself; and surely outer space and the moon will in due course be enlisted for this unbecoming function, probably as the dumping ground for rocket-borne radioactive refuse.

In our time, whole lakes are dying of industrial exhaust. The seemingly isolated races of Lapland and Tierra del Fuego find their foodstuffs riddled with methyl mercury or radioactivity and must appeal to civilized societies to rescue them from their plight. The Atmospheric Sciences Research Center of Scotia, New York, reported in December 1969 that there was no longer a breath of uncontaminated air to be found anywhere in the North American hemisphere and predicted the universal use of artificial respirators throughout America within two decades. Thor Heyerdahl, sailing the Atlantic on the RA II expedition in 1970, reported finding not one oil-free stretch of water during the crossing. Jacques Piccard, exploring the depths of the seas, warned the United Nations in October 1971 that the oceans of the world would soon be incapable of sustaining aquatic life due to lead exhaust, oil dumping, and mercury pollution, with the Baltic, Adriatic, and Mediterranean seas already too far deteriorated to be saved.

But these now well-publicized forms of pollution are not the only distortive force the artificial environment exerts upon the rest of the world for the sake of sustaining its lifestyle. A single oil pipeline across the wild Alaskan tundra is enough to subordinate its entire ecology (ruinously) to urban-industrial needs. A single superhighway built from São Paulo to Brasilia deprives an entire rainforest of its autonomy. Already the land bordering the Trans-Amazonian Highway has been staked out for commercial and urban development; the beasts are being killed or driven off and the natives coerced into compliance with official policy by methods that include the strategic use of infectious diseases. The fact is, there remains little wilderness anywhere that does not have its resources scheduled on somebody's industrial or real estate agenda, less still that is not already piped and wired through with the city's necessities or criss-crossed by air traffic skylanes.

And then there is the tourism that goes out from the cities of the affluent societies like a non-stop attack of locusts. Whatever outright industrial pollution and development may spare, tourism—now the world's largest money-making industry—claims for its omnivorous appetite. There are few governments that have the stamina and self-respect to hold out against the brutal pressure to turn their land and folkways into a commercial fraud for the opulent foreigners who flatter themselves that they are "seeing the world." All the globe-trotters really see, of course, (or want to see), is a bit of commercialized ethnic hokum and some make-believe wilderness. Just as the world becomes the dumping ground of the urban-industrial societies, it also becomes their amusement park. And how many are there now, even among my readers, assiduously saving up for summer safaris in Kenya or whirlwind junkets of "the enchanted Orient," without any idea what a destructive entertainment they are planning—but of course at bargain prices?

The remnants of the natural world that survive in the experience of urban-industrial populations—like the national parks we must drive miles to see, only to find them cluttered with automobiles, beer cans, and transistor radios—are fast becoming only a different order of artificiality, islands of carefully doctored wilderness put on display for vacationers and boasting all the comforts of gracious suburban living. It is hard to imagine that within another few generations the globe will possess a single wild area that will be more than 30 minutes removed by helicopter from a television set, an air-conditioned deluxe hotel, and a Coca-Cola machine. By then, the remotest regions may well have been staked out for exotic tours whose price includes the opportunity to shoot a tiger or harpoon a whale as a souvenir of one's rugged vacation adventure. The natives will be flown in from central casting and the local color will be under the direction of Walt Disney productions. The visitors—knowing no better—will conceive of this charade as "getting away to nature." But in truth it will be only another, and a climactic aspect of the urban-industrial expansion.

What we have here is an exercise in arrogance that breaks with the human past as dramatically and violently as our astronauts in their space rockets break from the gravitational grip of the earth. And the destination toward which we move is already clearly before us in the image of the astronaut. Here we have man encapsulated in a *wholly* man-made environment, sealed up and surviving securely in a plastic womb that leaves nothing to chance or natural process. Nothing "irrational" meaning nothing man has not made, or made allowance for can intrude upon the astronaut's life space. He interacts with the world beyond his metallic epidermis only by way of electronic equipment; even his wastes are stored up within his self-contained, mechanical envelope. As for the astronaut himself, he is almost invariably a military man. How significant it is that so much of our future, both as it appears in science fiction and as it emerges in science fact, should be dominated by soldiers—the most machine-tooled and psychically regimented breed of human being: men programmed and under control from within as from without. Can any of us even imagine a future for urban-industrial society in which the heroes and leaders—those who explore the stars and handle the crises—are not such a breed of warrior-technician?

What is there left of the human being in our militarized space programs but a small knot of neural complexity not yet simulable by electronic means, obediently serving the great technical project at hand by integrating itself totally with the apparatus surrounding it? In this form—cushioned and isolated within a prefabricated, homeostatic life space and disciplined to the demands of the mechanisms which sustain it—the astronaut perfects the artificial environment. Here is a human being who may travel anywhere and say, "I am not part of this place or that. I am autonomous. I make my own world after my own image." He is packaged for export anywhere in the universe. But ultimately all places become the same gleaming, antiseptic, electronic, man-made place, endlessly reproduced. Ambitious "world-planners," like the students of Buckminster Fuller, already foresee a global system of transportable geodesic domes that will provide a standardized environment in every quarter of the earth. Something of such a world is with us now in the glass-box architecture of our jet-age airports and high-rise apartments. One can traverse half the earth in passing from one such building to another, only to discover oneself in a structure indistinguishable from that which one has left. Even the piped-in music is the same.

These are momentous developments. The astronautical image of man—and it is nothing but the quintessence of urban-industrial society's pursuit of the wholly controlled, wholly artificial environment—amounts to a spiritual revolution. This is man as he has never lived before; it draws a line through human history that almost assumes the dimensions of an evolutionary turning point. So it has been identified by Teilhard de Chardin, who has given us the concept of the "noosphere," a level of existence that is to be permanently dominated by human intellect and planning, and to which our species must now adapt if it is to fulfill its destiny. So, too, Victor Ferkiss has described technological man as a creature on the brink of an "evolutionary breakthrough." Technology, by giving man "almost infinite power to change his world and to change himself," has ushered in what Ferkiss calls an "existential revolution" whose spirit is summarized by the words of Emmanuel Mesthene:

"We have now, or know how to acquire, the technical capability to do very nearly anything we want. Can we transplant hearts, control personality, order the weather that suits us, travel to Mars or Venus? Of course we can, if not now or in five years or ten years, then certainly in 25 or in 50 or in 100."

The Greek tragedians would have referred to such a declaration as *hubris*—the overweening pride of the doomed. It remains *hubris*; but its moral edge becomes blunted as the sentiment descends into a journalistic cliché. Moreover, we have no Sophoclean operations analyst to give us a cost-benefit appraisal of its spiritual implications. The sensibility that accompanies technological omnipotence lacks the tragic dimension; it does not take seriously the terrible possibility that a society wielding such inordinate power may release reactive forces within the human psyche, as well as within the repressed natural environment, that will never allow it to survive for the 50 or 100 years it needs to exploit its capabilities....

Our politics has become deeply psychological, a confrontation of sanities. But if our psychology is not itself to be debased by scientific objectification, then it must follow where liberated consciousness leads it; into the province of the dream, the myth, the visionary rapture, the sacramental sense of reality, the transcendent symbol. Psychology, we must remember, is the study of the soul, therefore the discipline closest to the religious life. An authentic psychology discards none of the insights gained from spiritual disciplines. It does not turn them into a scholarly boneyard for reductive "interpretations," or regard them as an exotic and antiquated mysticism. Rather, it works to reclaim them as the basis for a rhapsodic intellect which will be with us always as a normal part of our common life.

And suppose the reality we live by should experience such a revolution ... what sort of political program would follow from that?

Nothing less, I think, than that we should undertake to repeal urban-industrialism as the world's dominant style of life. We should do this, not in a spirit of grim sacrifice, but in the conviction that the reality we want most to reside in lies beyond the artificial environment. And so we should move freely and in delight toward the true postindustrialism: a world awakened from its sick infatuation with power, growth, efficiency, progress as if from a nightmare.

The Parable of the Tribes: The Problem of Power in Social Evolution (1995)

Andrew Bard Schmookler

The ecosystem has been changed by civilization. To be sure, the old natural structures remain recognizable on the terrestrial landscape. But with the power of civilization over nature steadily growing, the old structures are subverted and replaced at an accelerating pace.

The ancient and time-proven patterns of cooperation give way to a regime of domination. Where previously all were free though unwitting actors in a collective drama of mutual survival, with civilized man there arrived on the scene a single player to write the script for the whole. To secure a place in the old synergistic system, a life form had to serve the ecosystem as a whole. But, increasingly, the prerequisite for a continuing role in the drama of life is service to the single dominant animal. If you impinge upon the interests of man, out you go: wolves and bears and lions, who like the meat that man wants for himself, are eradicated or at best are forced to retreat to refuges. If you are useless to man, however teeming with life, you will be swept aside in favor of something that better serves the master: the magnificent forests are felled and replaced by the more paltry but more "useful" growths of man's cropland. The grains and cattle that fill men's bellies—these thrive and prosper. Life comes to be governed by a calculus that is fundamentally corrupt. The well-being of man is what rules, regardless of how small may be the human benefit in relation to the costs in well-being to others of God's creatures. Never before has a creature had the power to arrange the pattern of life for its selfish ends, so never before has the ecosystem been corrupted. So pervasive is the assumption of the human right to selfishness in the ecosystem—might makes right—that even the arguments for human restraint tend to be couched in terms of human self-interest: natural environments have recreational value; species we extinguish might have proved later to have unforeseen usefulness to man.

There is one more case for restraint based on enlightened self-interest. Just as synergy is nature's tool for long-term viability, so also the wages of corruption are the long-run decadence and death of living systems. Man uses up the bases of his life. Look at civilization's most ancient homes: once fertile places, many of them now lie denuded of life's basic nutrients. Around the Mediterranean, across the "Fertile" Crescent, deforestation and overgrazing broke the grip by which the living system clung to the sacred soil. The spread of the deserts is accelerating. And for each bushel of corn that comes from Iowa, more than a bushel of its precious soil washes away. Man's corrupt pattern is feast and famine. In that order. The world's fisheries are overfished. The fragile forests of the tropics are recklessly harvested. Across the board, we take in for our use more than we or nature can replace. We have a strip miner's approach to our planet.

The decadence of civilization as a living system is demonstrated by the nonrecycling of its outputs as much as by the nonrenewal of its inputs. For every other living thing, its outputs function as essential inputs for others: the oxygen/carbon dioxide exchanges of plants and animals, the buildup of soil by the leaves that drop from trees and by the excrement of animals. Before civilization, life produced no toxic wastes. Now, our insecticides threaten birds and other species. Our burning of fuels may bring climatic disaster. Our output of fluorocarbons may expose us and other living things to harmful solar radiation. This generation is producing mountains of nuclear wastes that hundreds of generations to come will have to live with, and perhaps die from. And as frightening as radiation is, many warn us that we have more to fear from the countless tons of "conventional" wastes that lurk in thousands of dumpsites across the land. Out of the womb of civilization, down through countless Love Canals, issues forth death.

Civilization has shattered the intricate web that stabilizes the flows of life. Awareness of this problem has grown dramatically in just the past generation. But the direction of the biosphere's movement under the continuing impact of civilization is still toward degradation and decadence. So rapid is the growth and spread of civilization's power that the pace of death has, if anything, accelerated. All life is so interdependent that either we must stop the decline of the biosphere or fall with it, and we must be quick about it. Either quick, or dead....

The *intersocietal system* of civilization, as we have seen in Part I, is an arena for unregulated conflict. Civilization created conflict by opening for each civilized society possibilities that fostered conflicts of interest among societies, and by creating an anarchic situation that mitigated against synergistic action on the basis of interests shared by those societies. The consequent ceaseless struggle for power has been unsynergistic in several ways.

First, conflict gains its role in the intersocietal system even against the wishes of mankind. The parable of the tribes shows how even if all or almost all wish to live in peace and safety the structure of the intersocietal system prevents this optimal condition from prevailing. As the general historic plague of war comes to mankind uninvited, so too there occur specific wars no one wanted and other wars that whether wanted or not, benefit no one. As I write, a war is ongoing between Iraq and Iran of which it has been said, "It is a war both sides are losing."

Second, even when some benefit from the conflict, the struggle for power is almost invariably a minus-sum game, one in which the net gains of the winners are more than offset by the net losses of the losers. War is costly to wage, and the destruction wrought by it leaves the whole less than it was at the outset. But beyond those factors is a more important one akin to the economic idea of diminishing marginal utility: in most human affairs the movement from some to much gives less benefit than the movement from none to some. It follows that the movement from some to none does more harm than the movement from some to much does good. Thus the conqueror who now governs two lands may be better off, but his gain is not commensurate with the loss of the vanquished who is dispossessed. The profit of gaining a slave is far less than the debit of losing one's liberty. Yet, the history of civilization is full of just such exchanges imposed by uncontrolled force. The pursuit of such conflict may be "rational" (in, again, the economic sense of the pursuit of self-interest) from the point of view of the stronger party who stands to gain, but it is irrational and unsynergistic from the point of view of the system as a whole. In natural systems, such choices do not arise, for the power to injure the whole for the sake of oneself is granted no one. The unprecedented anarchy of civilization's intersocietal system breaks down the order of synergy, making room for the corrupt regime of power.

Third, the immediate costs of the corrupt rule of power are compounded by the long-term social evolutionary costs. Out of the strife comes a selective process leading people along a path different from what they would have chosen. The absence of an overarching synergy to assure that intersocietal interactions serve the common interests has condemned mankind to domination by ever-escalating power systems largely indifferent to the well-being of human beings or other living creatures.

This unsynergistic determination of our social evolutionary destiny clearly endangers the long-term viability of the system. Never before has a living creature had in its repertoire of possible actions the virtual destruction of itself and other life on earth. Always, there might have streamed out of the indifferent heavens some giant meteor or comet or asteroid to burst the thin film of life's bubble on this planet. But living things, having been designed with no other options, always served life.... For the first time in more than three billion years of life, a living system is relentlessly creating the means not of self-preservation, but of self-destruction.

CRITIQUE OF CYNICAL REASON (1987)

PETER SLOTERDIJK

Cynicism: The Twilight of False Consciousness

And indeed no longer was anyone to be seen who stood behind everything. Everything turned continually about itself. Interests changed from hour to hour. Nowhere was there a goal anymore.... The leaders lost their heads. They were drained to the dregs and calcified.... Everyone in the land began to notice that things didn't work anymore.... Postponing the collapse left one path open.

—Franz Jung, *Die Eroberung der Maschinen* (1921)

The discontent in our culture has assumed a new quality: It appears as a universal, diffuse cynicism. The traditional critique of ideology stands at a loss before this cynicism. It does not know what button to push in this cynically keen consciousness to get enlightenment going. Modern cynicism presents itself as that state of consciousness that follows after naive ideologies and their enlightenment. In it, the obvious exhaustion of ideology critique has its real ground. This critique has remained more naive than the consciousness it wanted to expose; in its well-mannered rationality, it did not keep up with the twists and turns of modern consciousness to a cunning multiple realism. The formal sequence of false consciousness up to now—lies, errors, ideology—is incomplete; the current mentality requires the addition of a fourth structure: the phenomenon of cynicism. To speak of cynicism means trying to enter the old building of ideology critique through a new entrance.

It violates normal usage to describe cynicism as a universal and diffuse phenomenon; as it is commonly conceived, cynicism is not diffuse but striking, not universal but peripheral and highly individual. The unusual epithets describe something of its new manifestation, which renders it both explosive and unassailable....

The fertile ground for cynicism in modern times is to be found not only in urban culture but also in the courtly sphere.

Both are dies of pernicious realism through which human beings learn the crooked smile of open immorality. Here, as there, a sophisticated knowledge accumulates in informed, intelligent minds, a knowledge that moves elegantly back and forth between naked facts and conventional façades. From the very bottom, from the declassed, urban intelligentsia, and from the very top, from the summits of statesmanly consciousness, signals penetrate serious thinking, signals that provide evidence of a radical, ironic treatment (*Ironisierung*) of ethics and of social conventions, as if universal laws existed only for the stupid, while that fatally clever smile plays on the lips of those in the know. More precisely, it is the powerful who smile this way, while the cynical plebeians let out a satirical laugh. In the great hall of cynical knowledge the extremes meet: Eulenspiegel meets Richelieu; Machiavelli meets Rameau's nephew; the loud Condottieri of the Renaissance meet the elegant cynics of the rococo; unscrupulous entrepreneurs meet disillusioned outsiders; and jaded systems strategists meet conscientious objectors without ideals.

Since bourgeois society began to build a bridge between the knowledge of those at the very top and those at the very bottom and announced its ambition to ground its worldview completely on *realism*, the extremes have dissolved into each other. Today the cynic appears as a mass figure: an average social character in the upper echelons of the elevated superstructure. It is a mass figure not only because advanced industrial civilization produces the bitter loner as a mass phenomenon. Rather, the cities themselves have become diffuse clumps whose power to create generally accepted *public characters* has been lost. The pressure toward individualization has lessened in the modern urban and media climate. Thus modern cynics—and there have been mass numbers of them in Germany, especially since the First World War—are no longer outsiders. But less than ever do they appear as a tangibly developed type. Modern mass cynics lose their individual sting and refrain from the risk of letting themselves be put on display. They have long since ceased to expose themselves as eccentrics to the attention and mockery of others. The person with the clear, "evil gaze" has disappeared into the crowd; anonymity now becomes the domain for cynical deviation. Modern cynics are integrated, asocial characters who, on the score of subliminal illusionlessness, are a match for any hippie. They do not see their clear, evil gaze as a personal defect or an amoral quirk that needs to be privately justified. Instinctively, they no longer understand their way of existing as something that has to do with being evil, but as participation in a collective, realistically attuned way of seeing things. It is the universally widespread way in which enlightened people see to it that they are not taken for suckers. There even seems to be something healthy in this attitude, which, after all, the will to self-preservation generally supports. It is the stance of people who realize that the times of naiveté are gone.

Psychologically, present-day cynics can be understood as borderline melancholics, who can keep their symptoms of depression under control and can remain more or less able to work. Indeed, this is the essential point in modern cynicism; the ability of its bearers to work—in spite of anything that might happen, and especially, after anything that might happen. The key social positions in boards, parliaments, commissions, executive councils, *publishing companies*, practices, faculties, and lawyers' and editors' offices have long since become a part of this diffuse cynicism. A certain chic bitterness provides an undertone to its activity. For cynics are not dumb, and every now and then they certainly see the nothingness to which everything leads. Their psychic (*seelisch*) apparatus has become elastic enough to incorporate as a survival factor a permanent doubt about their own activities. They know what they are doing, but they do it because, in the short run, the force of circumstances and the instinct for self-preservation are speaking the same language, and they are telling them that it has to be so. Others would do it anyway, perhaps worse. Thus, the new, integrated cynicism even has the understandable feeling about itself of being a victim and of making sacrifices. Behind the capable, collaborative, hard façade, it covers up a mass of offensive unhappiness and the need to cry. In this, there is something of the mourning for a "lost innocence," of the mourning for better knowledge, against which all action and labor are directed.

THE SEEDS OF TIME (1994)

FREDRIC JAMESON

The paradox from which we must set forth is the equivalence between an unparalleled rate of change on all the levels of social life and an unparalleled standardization of everything—feelings along with consumer goods, language along with built space—that would seem incompatible with just such mutability. It is a paradox that can still be conceptualized, but in inverse ratios: that of modularity, for example, where intensified change is enabled by standardization itself, where prefabricated modules, everywhere from the media to a henceforth standardized private life, from commodified nature to uniformity of equipment, allow miraculous rebuildings to succeed each other at will, as in fractal video. The module would then constitute the new form of the object (the new result of reification) in an informational universe: that Kantian point in which raw material is suddenly organized by categories into an appropriate unit.

But the paradox can also incite us to rethink our conception of change itself. If absolute change in our society is best represented by the rapid turnover in storefronts, prompting the philosophical question as to what has really changed when video stores are replaced by T-shirt shops, then Barthe's structural formulation comes to have much to recommend it, namely, that it is crucial to distinguish between rhythms of change inherent to the system and programmed by it, and a change that replaces one entire system by another one altogether. But that is a point of view that revives paradoxes of Zeno's sort, which derive from the Parmenidean conception of Being itself, which, as it is by definition, cannot be thought of as even momentarily becoming, let alone failing to be for the slightest instant.

The "solution" to this particular paradox lies of course in the realization (strongly insisted on by Althusser and his disciples) that each system—better still, each "mode of production"—produces a temporality that is specific to it: it is only if we adopt a Kantian and ahistorical view of time as some absolute and empty category that the peculiarly repetitive temporality of our own system can become an object of puzzlement and lead to the reformulation of these old logical and ontological paradoxes.

Yet it may not be without its therapeutic effects to continue for one long moment to be mesmerized by the vision attributed to Parmenides, which however little it holds for nature might well be thought to capture a certain truth of our social and historical moment: a gleaming science-fictional stasis in which appearances (simulacra) arise and decay ceaselessly, without the momentous stasis of everything that is flickering for the briefest of instants or even momentarily wavering in its ontological prestige.

Here, it is as if the logic of fashion had, accompanying the multifarious penetration of its omnipresent images, begun to bind and identify itself with the social and psychic fabric in some ultimately inextricable way, which tends to make it over into the very logic of our system as a whole. The experience and the value of perpetual change thereby comes to govern language and feelings, fully as much as the buildings and the garments of this particular society, to the point at which even the relative meaning allowed by uneven development (or "nonsynchronous synchronicity") is no longer comprehensible, and the supreme value of the New and of innovation, as both modernism and modernization grasped it, fades away against a steady stream of momentum and variation that at some outer limit seems stable and motionless.

What then dawns is the realization that no society has ever been so standardized as this one, and that the stream of human, social, and historical temporality has never flowed quite so homogeneously. Even the great boredom or ennui of classical modernism required some vantage point or fantasy subject position outside the system; yet our seasons are of the post-natural and postastronomical television or media variety, triumphantly artificial by way of the power of their National Geographic or Weather Channel images: so that their great rotations—in sports, new model cars, fashion, television, the school year or *rentrée*, etc.—stimulate formerly natural rhythms for commercial convenience and reinvent such archaic categories as the week, the month, the year imperceptibly, without any of the freshness and violence of, say, the innovations of the French revolutionary calendar.

What we now begin to feel, therefore—and what begins to emerge as some deeper and more fundamental constitution of postmodernity itself, at least in its temporal dimension—is that henceforth, where everything now submits to the perpetual change of fashion and media image, nothing can change any longer. This is the sense of the revival of that "end of History" Alexandre Kojéve thought he could find in Hegel and Marx, and which he took to mean some ultimate achievement of democratic equality (and the value equivalence of individual economic and juridical subjects) in both American capitalism and Soviet communism, only later identifying a significant variant of it in what he called Japanese "snobisme," but that we can today identify as postmodernity itself (the free play of masks and roles without content or substance). In another sense, of course, this is simply the old "end of ideology" with a vengeance, and cynically plays on the waning of collective hope in a particularly conservative market climate. But the end of History is also the final form of the temporal paradoxes we have tried to dramatize here: namely, that a rhetoric of absolute change (or "permanent revolution" in some trendy and meretricious new sense), is, for the postmodern, no more satisfactory (but not less so) than the language of absolute identity and unchanging standardization cooked up by the great corporations, whose concept of innovation is best illustrated by the neologism and the logo and their equivalents in the realm of built space, "lifestyle," corporate culture, and psychic programming. The persistence of the Same through absolute Difference—the same street with different buildings, the same culture through momentous new sheddings of skin—discredits change, since henceforth the only conceivable radical change would consist in putting an end to change itself. But here the antinomy really does result in the blocking or paralysis of thought, since the impossibility of thinking another system except by way of the cancellation of this one ends up discrediting the Utopian imagination itself, which is fantasized, as we shall see later, as the loss of everything we know experientially, from our libidinal investments to our psychic habits, and in particular the artificial excitements of consumption and fashion.

"THE MEDIUM IS THE MEDIUM" (1998)
THERE IS NO EQUAL

LABOR OF LUDD

borigines anticipate apocalypse ... agriculture aggrandizes arable areas and allots acreage, assuming acquisition and alienation ... arithmetic adds another abstract axis ... authority appreciates art—already accepting abstractions' ascendancy—as authenticating appearances ... by banishing bounty, bureaucracy's blackmail breeds bitterness between brothers behind benign banality; business believes boundless buying brings back bliss ... commodity circulation controls current conditions completely, calculating career compulsions can continue consumption, constantly creating cruel contradictions colonized consciousness conveniently corrects ... dreams distill dormant desires, darkly divining domestication's demise ... disrupting digital discourse dialectically demonstrates *dash*, dooming domination's designer discipline ... duplicity defeats double-driveling duplication ... equations empower everyday economics, essentially encoding estranged enterprise; elegant ecstasy ebbs ... "environment" equals earth? ... formula for fusing *formally* fragments freed from function's foundation: fully further facsimiles' fulfillment; feature "forbidden" fantasies fully filmed; finally, fabricate fetishes fascinating feelings for fashion ... grammar guards God's grave ... hell, having had heaven's hallucinatory holiday haunting hearts held history's hostage has hardly helped humanist hacks humble humanity's heretical haughtiness ... images interpose intermediating influences inside *interests*; insubordination is interested in insinuating illusion into identifying itself ... insolence insists its intelligence is inimitably incendiary, illuminating irony's impotence ... jaded judges jeopardize justice ... know krime kan konjure komedy kontaining kommunist kontent ... lush laughing lust launches life; lavishly littered *likenesses*, like, lessen life's lure ... language licenses lucidity logically; licentious lucidity loosens letters' lock laughingly, luminously liquidating leaden logic ... languorous looting lampoons leisure ... modestly managing *mas(s) o' (s)chism(s)* mutilates multitudes ... matchless money makes mastery meaningless: modern mutiny must make meaning menace mediation: mimicry means *mirror's measure matched* ... nowadays nihilism's nothing new ... our offense? outwitting our overseers' overly optimistic overthrow of our original *obliquity* ... private property produces parity—parity portends production's ponderous planet-punishing progress piss-pure puns parody preyfully ... quality's quintessence quickens ... relentlessly replicating reality ripens revolts rigorously resisting representations' recuperations; rewinding reality readies really radical reversals ... school separates subjects, subjecting subjectivity so separations seem sane ... scholastic scavengers scrutinize signs showing *signification* scarcely sustains synthetic scarcity ... theory that threatens to transform the totality transgresses tedium; tongue-twisters tend to turn topsy-turvy the tyranny that *things talking to themselves* typifies ... the training that teaches those throngs to trade themselves to time trembles ... ultimately, understanding urban upsurges' unconscious urges uncovers undercurrents undermining uncannily *utility*'s ugly unwitting velocity ... videos vacuous veneer veils vast vulgarities: vanishing vitality, vehement veracity, vapid vanity ... we wage war with words, wither wage work's wearying world whenever we wield wit which wickedly widens wild wholeness while working wonders ... xorcising xiled xistence's xtraordinary xhaustion xposes xchange, ... your yoke yields yet you yawn ... z z z z z.

"How Nice to Be Civilized!" (1993)

Des Réfractaires

Assassinations, massacres, rape, torture: these crimes committed on the soil of what was once Yugoslavia are not the acts of uncontrollable savages; of educationless brutes.

No doubt as children they respected the family order; are now more or less faithful followers of religions; earnest sports spectators; content with television. In a word, civilized folks; normal people doing what society expects them to!

Each crime demonstrates the success of diverse processes of domestication which have come to be grouped under the heading of Civilization.

The killers, rapists and perpetrators of massacres have exceptionally well internalized today's world's fundamental logic: to survive, other people must be destroyed! This mutual mangling takes different forms, such as economic competition or war. But the result is always the same: some must be trampled in order to give others the impression that they are living more and better. Being civilized signifies not taking your own life and those of others into consideration. It means letting your life be used, exploited and dominated by the always-superior interests of the collectivity where fate decreed that you would be born and live your life. And all for the financial, etc., gain of the authorities of the collectivity in question. In exchange for this submission one is granted the possibility of being accepted as a human being.

Being civilized, as well, signifies sacrificing your life, and those of others, when those in power attempt to solve their management problems with wars.

Aside from a variety of benefits they offer, wars represent a very efficient means of directing feelings of frustration against people who, designated as prey, can then be oppressed, humiliated and killed without qualms. Those who suffer, as with those who take pleasure in making others suffer, become nothing more than instruments of the conditions of social existence, conditions where lives are only important in relation to the use that can be made of them.

Following the collapse and decomposition of the Eastern Bloc, various local and international gangsters have slots to fill, markets to conquer and energies to channel through the formation of new States.

To help slice up the pie, local political gangs have deftly played the religious and nationalist cards. And if these cards work effectively, unfortunately, it is because, for a portion of the population, this collapse and decomposition have not been perceived as openings towards increased freedom. On the contrary, people have experienced an immense emptiness, one that has been alleviated with nationalist and religious alienations which are often decked out in a tawdry grab-bag of local history and culture. Instead of attempting to understand and attack the real causes of our material and psychological misery, too often people are thrown into a state of disarray. In response to this disarray identities are presented as lost values to be recaptured, whereas these values are simply the ideological cement which is the prerequisite to founding and developing State entities propped up by alliances between local and world powers.

Nor, in a climate of generalized terror, is there any hesitation to accomplish this by displacing populations and practicing ethnic cleansing in order to redistribute land. In this sense, don't the peace plan concocted in Geneva and hypothetical military intervention rubberstamp the UN's recognition of the dismemberment of the territory of former Yugoslavia? And if this is to be the price of pacification, everyone just closes their eyes to the cortege of horrors which is integral to every war.

The humanitarian organizations, cynically baptized non-governmental, present the dismal paradox of inciting pity and indignation while at the same time impeding the possibility of spontaneous participation from which true human solidarity could be born.

Today humanitarianism is a true lobby in a financial, human and media sense. But beyond generating money, humanitarianism carries out an educational task, channeling emotions and arousing feelings of indignation on a specific and regular basis—paving the way to military intervention in humanitarian wars which the State undertakes to supposedly respond to pressure from a public indignant about the very real massacre that they are powerlessly witnessing. This type of media treatment's only goal is to convince people that alone, by themselves, nothing can be done; the State is in a position to come to the rescue and will watch out for their political and strategic interests.

Everything is peachy because everyone consoles themselves with the thought that peace and democracy are a privilege—the proof being that elsewhere, over there, all is war and barbarism.

Denouncing the horrors, collecting accounts from the local population, exhorting the government to intervene, the media have the starring role in this affair. Real recruiting sergeants! As to be expected, the media have carefully edited out any information about those in ex-Yugoslavia who oppose the war, carefully concealing information about the 1992 massacres in Zagreb and Sarajevo which put the finishing touches on repressing the movements against the war. These horrors are necessary in order to lay the basis for the right to intervene, to invent humanitarian wars and to create tribunals to judge the vanquished. The "New World Order" which is coming into being is cutting its teeth on small nation-State wars; it provides the arms, then comes to the rescue, basing its activities in each case on a flood of horrifying images!

Thus exalting ethnic, national and religious identities goes hand in hand with gang warfare to constitute a new hierarchy of Godfathers.

In response to the growth of ghettos—those artificial separations and false communities which allow the world of money and domination to thrive on human life—we, as people who are refractory to the world around us, would like to affirm our community of struggle and aspirations with those who are refusing the war in ex-Yugoslavia, those who see themselves above all as "human beings who want to live" and not cannon fodder.

We are refractory to all that is the glory of civilization. We want to live human relations that would no longer be based on appropriation, competition and hierarchy, and would thus be relations in which individuals would no longer be obliged to treat themselves *a priori* as adversaries and enemies.

"Civilization in Bulk" (1991)

David Watson

aving had the privilege of living for a time among stone-age peoples of Brazil, a very civilized European of considerable erudition wrote afterwards, "Civilization is no longer a fragile flower, to be carefully preserved and reared with great difficulty here and there in sheltered corners.... All that is over: humanity has taken to monoculture, once and for all, and is preparing to produce civilization in bulk, as if it were sugar-beet. The same dish will be served to us every day."

Those words were written in 1955. Now that civilization is engulfing the entire planet, the image of the fragile flower has largely wilted. Some of civilization's inmates are remembering that the image was always a lie; other ways of seeing the world are being rediscovered. Counter-traditions are being reexamined, escape routes devised, weapons fashioned. To put it another way, a spectre haunts the heavy equipment as it chugs deeper into the morass it has made: the spectre of the primal world.

Devising escapes and weapons is no simple task: false starts and poor materials. The old paths are paved and the materials that come from the enemy's arsenal tend to explode in our hands. Memory and desire have been suppressed and deformed; we have all been inculcated in the Official History. Its name is Progress, and the Dream of Progress continues to fuel global civilization's expansion everywhere, converting human beings into mechanized, self-obliterating puppets, nature into dead statuary.

The Official History can be found in every child's official history text: Before the genesis (which is to say, before civilization), there was nothing but a vast, oceanic chaos, dark and terrible, brutish and nomadic, a bloody struggle for existence. Eventually, through great effort by a handful of men, some anonymous, some celebrated, humanity emerged from the slime, from trees, caves, tents and endless wanderings in a sparse and perilous desert to accomplish fantastic improvements in life. Such improvements came through mastery of animals, plants and minerals; the exploitation of hitherto neglected Resources; the fineries of high culture and religion; and the miracles of technics in the service of centralized authority.

This awe-inspiring panoply of marvels took shape under the aegis of the city-state and behind its fortified walls. Through millennia, civilization struggled to survive amid a storm of barbarism, resisting being swallowed by the howling wilderness. Then another "Great Leap Forward" occurred among certain elect and anointed kingdoms of what came to be called "the West," and the modern world was born: the enlightenment of scientific reason ushered in exploration and discovery of the wilderness, internal (psychic) and external (geographic). In the kingdom's official murals, the Discoverers appear at one end, standing proudly on their ships, telescopes and sextants in their hands; at the other end waits the world, a sleeping beauty ready to awake and join her powerful husband in the marriage bed of nature and reason.

Finally come the offspring of this revolution: invention, mechanization, industrialization, and ultimately scientific, social and political maturity, a mass democratic society and mass-produced abundance. Certainly, a few bugs remain to be worked out—ubiquitous contamination, runaway technology, starvation and war (mostly at the uncivilized "peripheries"), but civilization cherishes its challenges, and expects all such aberrations to be brought under control, rationalized through technique, redesigned to "serve human needs," forever and ever, amen. History is a gleaming locomotive running on rails—albeit around precarious curves and through some foreboding tunnels—to the Promised Land. And whatever the dangers, there can be no turning back.

A False Turn

But now that several generations have been raised on monoculture's gruel, civilization is coming to be regarded not as a promise yet to be fulfilled so much as a maladaption of the species, a false turn or a kind of fever threatening the planetary web of life. As one of History's gentle rebels once remarked, "We do not ride upon the railroad; it rides upon us." The current crisis, occurring on every level, from the ecospheric to the social to the personal, has become too manifest, too grievous, to ignore. The spectre haunting modern civilization, once only a sense of loss, now has open partisans who have undertaken the theoretical and practical critique of civilization.

So we begin by reexamining our list of chapters not from the point of view of the conquerors but the conquered: the slaves crushed under temple construction sites or gassed in the trenches, the dredged and shackled rivers, the flattened forests, the beings pinned to laboratory tables. What voice can better speak for them than the primal? Such a critique of "the modern world through Pleistocene eyes," such a "geological kind of perspective," as the indigenous authors of the 1977 Haudenosaunee (Iroquois) document, *A Basic Call to Consciousness*, put it, immediately explodes the conquerors Big Lie about "underdevelopment" and the "brutality" of primal society, their vilification of prehistory.

The lie has most recently been eroded not only by greater access to the views of primal peoples and their native descendants who are presently fighting for survival, but by a more critical, non-eurocentric anthropology willing to challenge its own history, premises and privilege. Primal society, with its myriad variations, is the common heritage of all peoples. From it, we can infer how human beings lived some 99 percent of our existence as a species. (And even a large part of that last one percent consists of the experience of tribal and other vernacular communities that resist conquest and control in creative, if idiosyncratic ways.)

Looking with new/old eyes on the primal world, we see a web of autonomous societies, splendidly diverse but sharing certain characteristics. Primal society has been called "the original affluent society," affluent because its needs are few, all its desires are easily met. Its tool kit is elegant and lightweight, its outlook linguistically complex and conceptually profound yet simple and accessible to all. Its culture is expansive and ecstatic. It is propertyless and communal, egalitarian and cooperative. Like nature, it is essentially leaderless: neither patriarchal nor matriarchal, it is anarchic, which is to say that no *archon* or ruler has *built* and occupied center stage. It is, rather, an organic constellation of persons, each unique.

A Society Free of Work

It is also a society free of *work*; it has no economy or production per se, except for gift exchange and a kind of ritual play that also happen to create subsistence (though it is a society capable of experiencing occasional hunger without losing its spiritual bearings, even sometimes *choosing* hunger to enhance interrelatedness, to play or to see visions). The Haudenosaunee, for example, write that "[we] do not have specific economic institutions, nor do we have specifically distinct political institutions." Furthermore, the subsistence activities of Haudenosaunee society, "by our cultural definition, [are] not an economy at all."

Hence, primal society's plenitude resides in its many symbolic, personal, and natural relationships, not in artifacts. It is a dancing society, a singing society, a celebrating society, a dreaming society. Its philosophy and practice of what is called animism—a mythopoetic articulation of the organic unity of life discovered only recently by the West's ecologists—protects the land by treating its multiplicity of forms as sacred beings, each with its own integrity and subjectivity. Primal society affirms community with all of the natural and social world.

Somehow this primal world, a world (as Lewis Mumford has observed) more or less corresponding to the ancient vision of the Golden Age, unravels as the institutions of kingship and class society emerge. How it happened remains unclear to us today. Perhaps we will never fully understand the mystery of that original mutation from egalitarian to state society. Certainly, no standard explanations are adequate. "That radical discontinuity," in the words of Pierre Clastres, "that mysterious emergence—irreversible, fatal to primitive societies— of the thing we know by the name of the State," how does it occur?

Primal society maintained its equilibrium and its egalitarianism because it refused power, refused property. Kingship could not have emerged from the chief because the chief had no power over others. Clastres insists: "Primitive society is the place where separate power is refused, because the society itself, and not the chief, is the real locus of power."

It is possible that we could approach this dissolution of original community appropriately only by way of mythic language like the Old Ones would have used. After all, only a poetic story could vividly express such a tragic loss of equilibrium. The latent potentiality for power and technique to emerge as separate domains had been previously kept at bay by the gift cycle, "techniques of the sacred" and the high level of individuation of society's members.

Primal peoples, according to Clastres, "had a very early premonition that power's transcendence conceals a mortal risk for the group, that the principle of an authority which is external and the creator of its own legality is a challenge to culture itself. It is the intuition of this threat that determined the depth of their political philosophy. For, on discovering the great affinity of power and nature, as the twofold limitation on the domain of culture, Indian societies were able to create a means for neutralizing the virulence of political authority."

This, in effect, is the same process by which primal peoples neutralized the potential virulence of technique: they minimized the relative weight of instrumental or practical techniques and expanded the importance of techniques of *seeing*: ecstatic techniques.... The shaman is, in Jerome Rotherberg's words, a "technician" of ecstasy, a "protopoet" whose "technique hinges on the creation of special linguistic circumstances, i.e., of song and invocation." Technology, like power, is in such a way refused by the dynamic of primal social relations. But when technique and power emerge as separate functions rather than as strands inextricably woven into the fabric of society, everything starts to come apart. "The unintended excrescence that grows out of human communities and then liquidates them," as Fredy Perlman called it, makes its appearance. A sorcery run amok, a golem-like thingness that outlives its fabricators: somehow the gift cycle is ruptured; the hoop, the circle, broken.

The community, as Clastres puts it, "has ceased to exorcise the thing that will be its ruin: power and respect for power." A kind of revolution, or counter-revolution, takes place: "When, in primitive society, the economic dynamic lends itself to definition as a distinct and autonomous domain, when the activity of production becomes alienated, accountable labor, levied by men who will enjoy the fruits of that labor, what has come to pass is that society has been divided into rulers and ruled, masters and subjects.... The political relation of power precedes and founds the economic relation of exploitation. Alienation is political before it is economic; power precedes labor; the economic derives from the political; the emergence of the State determines the advent of classes."

The emergence of authority, production and technology are all moments within the same process. Previously, power resided in no separate sphere, but rather within the circle—a circle that included the human community and nature (nonhuman kin). "Production" and the "economic" were undivided as well; they were embedded in the circle through gift sharing which transcends and neutralizes the artifactuality or "thingness" of the objects passing from person to person. (Animals, plants and natural objects being *persons*, even kin, subsistence is therefore neither work nor production, but rather gift, drama, reverence, reverie.) Technique also had to be embedded in relations between kin, and thus open, participatory, and accessible to all; or it was entirely personal, singular, visionary, unique and untransferable.

Equilibrium Exploded

The "great affinity of power and nature," as Clastres puts it, explains the deep cleft between them when power divides and polarizes the community. For the primal community, to follow Mircea Eliade's reasoning, "The world is at once 'open' and mysterious.... 'Nature' at once unveils and 'camouflages' the 'supernatural' [which] constitutes the basic and unfathomable mystery of the World." Mythic consciousness apprehends and intervenes in the world, participates in it, but this does not necessitate a relation of domination; it "does not mean that one has transformed [cosmic realities] into 'objects of knowledge.' These realities still keep their original ontological condition."

The trauma of disequilibrium exploded what contemporary pagan feminists have called "power within" and generated "power over." What were once mutualities became hierarchies. In this transformation, gift exchange disappears; gift exchange with nature disappears with it. What was shared is now hoarded: the mystery to which one once surrendered now becomes a territory to be conquered. All stories of the origins become histories of the origins... of the Master. The origin of the World is retold as the origin of the State.

Woman, who through the birth process exemplifies all of nature and who maintains life processes through her daily activities of nurturance of plants, animals and children, is suppressed by the new transformer-hero. Male power, attempting to rival the fecundity of woman, simulates birth and nature's fecundity through the manufacture of artifacts and monuments. The womb—a primordial container, a basket or bowl—is reconstituted by power into the city walls.

"Thus," as Frederick W. Turner puts it in *Beyond Geography: The Western Spirit Against the Wilderness*, "the 'rise to civilization' might be seen not so much as the triumph of a progressive portion of the race over its lowly, nature-bound origins as a severe, aggressive *volte-face* against all unimproved nature, the echoes of which would still be sounding millennia later when civilized men once again encountered the challenges of the wilderness beyond their city walls."

No explanation and no speculation can encompass the series of events that burst community and generated class society and the state. But the result is relatively clear: the institutionalization of hierarchic elites and the drudgery of the dispossessed to support them, monoculture to feed their armed gangs, the organization of society into work battalions, hoarding, taxation and economic relations, and the reduction of the organic community to lifeless resources to be mined and manipulated by the archon and his institutions.

The "chief features" of this new state society, writes Mumford, "constant in varying proportions throughout history, are the centralization of political power, the separation of classes, the lifetime division of labor, the mechanization of production, the magnification of military power, the economic exploitation of the weak, and the universal introduction of slavery and forced labor for both industrial and military purposes." In other words, a *megamachine* made up of two major arms, a labor machine and a military machine.

The crystallization of a fluid, organic community into a pseudo-community, a giant machine, was in fact the first machine, the standard definition of which, Mumford notes, is "a combination of resistant parts, each specialized in function, operating under human control, to utilize energy and perform work...." Thus, he argues, "The two poles of civilization then, are mechanically-organized work and mechanically-organized destruction and extermination. Roughly the same forces and the same methods of operation [are] applicable to both areas." In Mumford's view, the greatest legacy of this system has been "the myth of the machine"—the belief that it is both irresistible and ultimately beneficial. This mechanization of human beings, he writes, "had long preceded the mechanization of their working instruments.... But once conceived, this new mechanism spread rapidly, not just by being imitated in self-defense, but by being forcefully imposed...."

One can see the differences here between the kind of technics embedded in an egalitarian society and technics-as-power or technology. As Mumford argues, people "of ordinary capacity, relying on muscle power and traditional skills alone, were capable of performing a wide variety of tasks, including pottery and manufacture and weaving, without any external direction or scientific guidance, beyond that available in the tradition of the local community. Not so with the megamachine. Only kings, aided by the discipline of astronomical science and supported by the sanctions of religion, had the capacity of assembling and directing the megamachine. This was an invisible structure composed of living, but rigid, human parts, each assigned to his special office, role, and task, to make possible the immense work-output and grand designs of this great collective organization."

Civilization as Gulag

In his intuitive history of the megamachine, Fredy Perlman describes how a Sumerian "Ensi" or overseer, lacking the rationalizations of the ideology of Progress which are routinely used to vaccinate us against our wildness, might see the newly issued colossus:

"He might think of it as a worm, a giant worm, not a living worm but a carcass of a worm, a monstrous cadaver, its body consisting of numerous segments, its skin pimpled with spears and wheels and other technological implements. He knows from his own experience that the entire carcass is brought to artificial life by the motions of the human beings trapped inside, the zeks who operate the springs and wheels, just as he knows that the cadaverous head is operated by a mere zek, the head zek."

It is no accident that Fredy chose the word *zek*, a word meaning gulag prisoner that he found in Solzhenitsyn's work. It was not only to emphasize that civilization has been a labor camp from its origins, but to illuminate the parallels between the ancient embryonic forms and the modern global work machine presently suffocating the earth. While the differences in magnitude and historical development are great enough to account for significant contrasts, essential elements shared by both systems—elements outlined above—position both civilizations in a polarity with primal community. At one extreme stands organic community: an organism, in the form of a circle, a web woven into the fabric of nature. At the other is civilization: no longer an organism but organic fragments reconstituted as a machine, an organization; no longer a circle but a rigid pyramid of crushing hierarchies; not a web but a grid expanding the territory of the inorganic.

According to official history, this grid is the natural outcome of an inevitable evolution. Thus natural history is not a multiverse of potentialities but rather a linear progression from Prometheus' theft of fire to the International Monetary Fund. A million and more years of species life experienced in organic communities are dismissed as a kind of waiting period in anticipation of the few thousand years of imperial grandeur to follow. The remaining primal societies, even now being dragged by the hair into civilization's orbit along its blood-drenched frontier, are dismissed as living fossils ("lacking in evolutionary promise," as one philosopher characterized them), awaiting their glorious inscription into the wondrous machine.

Thus, as Fredy Perlman argued, imperialism is far from being the last stage of civilization but is embedded in the earliest stages of the state and class society. So there is always a brutal frontier where there is empire and always empire where there is civilization. The instability and rapidity of change as well as the violence and destructiveness of the change both belie empire's claim to natural legitimacy, suggesting once more an evolutionary wrong turn, a profoundly widening disequilibrium.

The frontier expands along two intersecting axes, centrifugal and centripetal. In the words of Stanley Diamond, "Civilization originates in conquest abroad and repression at home. Each is an aspect of the other." Thus outwardly, empire is expressed geographically (northern Canada, Malaysia, the Amazon, etc.; the ocean bottoms, even outer space) and biospherically (disruption of weather and climate, vast chemical experiments on the air and water, elimination and simplification of ecosystems, genetic manipulation). But the process is replicated internally on the human spirit: every zek finds an empire in miniature "wired" to the very nervous system.

So, too, is repression naturalized, the permanent crisis in character and the authoritarian plague legitimated. It starts with frightened obedience to the archon or patriarch, then moves by way of projection to a violent, numbed refusal of the living subjectivity and integrity of the other—whether found in nature, in woman, or in conquered peoples.

At one end of the hierarchic pyramid stands unmitigated power; at the other, submission mingles with isolation, fragmentation and rage. All is justified, by the ideology of Progress—conquest and subjugation of peoples, ruin of lands and sacrifice zones for the empire, self-repression, mass addiction to imperial spoils, the materialization of culture. Ideology keeps the work and war machines operating.

Ultimately, this vortex brings about the complete objectification of nature. Every relationship is increasingly instrumentalized and technicized. Mechanization and industrialization have rapidly transformed the planet, exploding ecosystems and human communities with monoculture, industrial degradation and mass markets. The world now corresponds more closely to the prophetic warnings of primal peoples than to the hollow advertising claims of the industrial system: the plants disappearing and the animals dying, the soils denuded along with the human spirit, vast oceans poisoned, the very rain turned corrosive and deadly, human communities at war with one another over diminishing spoils—and all poised on the brink of an even greater annihilation at the push of a few buttons within reach of stunted, half-dead head-zeks in fortified bunkers. Civilization's railroad leads not only to ecocide, but to evolutionary suicide. Every empire lurches toward the oblivion it fabricates and will eventually be covered with sand. Can a world worth inhabiting survive the ruin that will be left?

MEMORIES AND VISIONS OF PARADISE (1995)

RICHARD HEINBERG

In the last few years I have come to see that the economic and social foundations of civilization are inherently corrupt and corrupting. Only through a monumental act of insensitivity can one ignore the anguish of the native peoples of the world, who have endured 500 years (or more) of uninterrupted pillage and oppression at the hands of civilized conquerors. And in many respects the situation only seems to be getting worse. Recently the U.S. Congress approved a global trade agreement—GATT—that creates a nondemocratic de facto world government whose reins rest securely in the hands of huge and unimaginably wealthy transnational corporations, an agreement that promises to inflict vastly increased economic hardship on indigenous peoples everywhere. The forces of centralization and power have succeeded beyond their wildest dreams: the entire planet is becoming one great marketplace, with every last tree and stream, and the labor of nearly every human being, available to the highest bidder. Industrial civilization is invading every last corner of the globe, foreclosing every alternative, narrowing our options to two: participate or die. But participation is death, too. As the global population increases, as wealth and power become more concentrated, and as resources, habitats and species disappear, disaster looms.

Perhaps I did indeed sound too optimistic a note back in 1989. But my real point was not that Paradise is just around the corner. In fact, even then I believed that, if many current trends continue unabated, the next century is likely to be one of unprecedented horror and suffering for billions of people and for the rest of Nature as well. My point then and now, rather, is that this devastation is not the inevitable outworking of human nature. It reflects neither our origin nor our ultimate destiny, which I take to be no more sinister than those of any other creature on this planet.

It seems to me that we human beings, and particularly we civilized humans, are wounded and sick. We reproduce catastrophe because we ourselves are traumatized—both as a species and individually, beginning at birth. Because we are wounded, we have put up psychic defenses against reality and have become so cut off from direct participation in the multidimensional wildness in which we are embedded that all we can do is to navigate our way cautiously through a humanly designed day-to-day substitute world of symbols—a world of dollars, minutes, numbers, images, and words that are constantly being manipulated to wring the most possible profit from every conceivable circumstance. The body and spirit both rebel.

Yet we together—or any one of us—can in principle return at any time to our true nature, wild, whole, and free. This, it seems to me, has been the message of every true prophet. Whether through acquaintance with our "inner child," through meditation, through wordless play with small children or animals, or through a deep encounter with the wilderness, we can choose to activate the part of ourselves that still remembers how to feel, love, and wonder. Yes, we have a lot of work ahead and a lot of minds to change before we can together create sustainable, diverse, decentralized cultures and leave behind oppression, racism, sexism, and economic parasitism. But that process becomes much easier when we share a sense of possibility, an assurance that we do not have to invent Paradise so much as to return to it; an assurance that at our core we are pure, brilliant, and innocent beings. Our task is not to create ever more elaborate global structures to enforce social and environmental justice (though I sympathize with the motives of people who work toward that end), but to strip away the artificiality that separates us from the magical simplicity that is our wild biotic birthright.

The Paradise myth continues to transform my vision. Some day, perhaps, human beings will be a blessing to the biosphere of this planet. I can imagine new, wild cultures in which people will put more emphasis on laughter and play than on power and possessions, in which our intellects will be engaged in the challenge of increasing the diversity of life rather than merely in finding new ways to exploit it. The path from here to there is likely to be a rocky one, but the longer we wait, the less chance we will have of traversing it successfully.

"They're Always Telling Me I'm Too Angry" (1995)

Chrystos

Especially when I mention land theft or rape or genocide
They go to therapy to understand themselves
pound anonymous pillows safely with a stranger
in the closed room of improper behavior
There is
no pillow I'm angry with
As far as I'm concerned I'm too tired to be angry enough
Angry that I can't go anyplace
without seeing demeaning images & outright lies about Indian people
I'm livid that we can't even keep the few pitiful acres we have left
if they happen to have uranium or copper or coal
Furious that I never feel safe alone on the streets
Angry that other People of Color
are sometimes as oppressive as whites
because whites taught them
everything they think they know about Indians
Riled that an Indian friend asked me why
I hang out with all those Black people
Angry with myself that I wasn't fast enough to say
Why do you hang out with all those damn white folks
Steaming mad that a million people in this country
which is no longer in a recession
have no place to live
while office buildings sit empty for years
Enraged that you can buy a submachine gun in Florida
about any other kind of gun any place you want
while the army & the cops amass more than enough weapons to kill
every person on earth
Furious that my cousin got shot in the head
& lives now barely able to say his name
I'm mad as hell at alcohol, crack & child abuse
I could easily kill several million random white folks
just to feel a little balance on this poor earth

But I've known since I was little that no matter how many
of us they kill
it's only ok for us to help them kill other brown folks
or to cheat each other or hate each other
or to buy stuff & imitate whiteness
or to act like our own people are the real problems
& we're above it all
This is the pillow I'm hitting without any repercussions
Angry that women are in therapy
while men have increased tenfold raping and murdering
Furious with child porn
the use of children to sell toilet paper & laundry soap
Spitting with rage at intolerance   starvation   waste   greed
all of which are reflected in myself despite my efforts
to seek balance
Boiling mad at my inadequacies & terror
raging that I'm still tortured by terrible nightmares
more than 20 years after I last saw the man
who raped my childhood into razors & nut houses
a man to whom nothing has happened or will happen
a man who did it to many other children
a man who my aunt handed me a picture of & said
This is when we were all such a happy family
though she knows what he did
a man whom even my closest friends tell me I shouldn't kill
They're wrong
Furious with the beaten parents who didn't protect me
because they didn't think I was worth it
or that they were
who beat me to shut me up
Enraged that the Black medical student was suspended
for punching out a white one who wore blackface to a party as a joke
Ha Ha it's so funny when you pretend to be one of us
Ha Ha we're not angry when you do any damn stupid thing you please
then punish us for our feelings in the matter
Ha Ha we love it when you buy your children fake tipis & headdresses
& books by whites of our stories with pictures of us
as pink charming savages
Ha Ha we're so happy you want to get rid of us so you can have all our stuff
& rename it & explain it & defame it
I'm enraged with every lying son of a turd

who takes our taxes to go to Bermuda & relax
after spending our money to murder whoever is
the current enemy & it's sometimes us
I'm spitting with rage that most of my friends can barely scramble by
I'm angry that I can't sleep   that I hate myself
that I can't write as well as I want
because I'm so damn angry I can't breathe
Furious that nobody else seems to be angry
& they don't want me to be either
Enraged at this whole sodden rotting mess they keep calling
civilization
as it poisons the air & the water & kills everyone in its way
which is so barbaric as to lock up its Elders
for the crime of not being able to care for themselves
which thinks of age as disease instead of wisdom
which persists in calling queers sick or depraved or immoral
despite the so-called separation of church & state
which doesn't exist
Red hot that I have to defend my anger
that sometimes I'm the nice one in comparison
to an even angrier woman
& then I'm treated with more respect
which demeans us both
I'm sick to death of blank eyes/zombies/nice girls
& lesbians who take drugs so they won't be depressed
as though depression is bad when it is a very rational
response to our lives
& I have spent my life living inside numbing depression
without drugs, gritting my teeth through another hour & resisting suicide
with my bare hands because I can't bear to let them win
when so many of my loved ones have blown their brains out in despair
I'm disgusted with drunks
& everybody who thinks
they've alive only to please themselves
even though some of them are my friends
I'd like to kill reality
which I don't understand
I want to blow up every stupid university
pretending that it is teaching something new
when all that's happening is that students are officially treated like fools
until they care only about a piece of paper

& whether I have a piece of paper or not
All the pieces of paper all the degrees are burning up in my anger
Everyone will have to face each other as human
I'm sick of everyone who asks
What do you do?
As though some corporate title or college bs
is an identity
I want to tie up all the white supremacists into crosses
set fire to their hatred
I want to fight back with every tendon of my weary body
run by a mind who remembers the toilet taste of jail food
knows the brutality of nut houses
arms that remember straitjackets & forced drugs & the screams
of women being dragged off to shock torture
knowing that to speak up too loudly means to be killed
because decent people
beat pillows or their wives instead of racism or hunger
because the idea of being nice is more important
than the idea of being real
It's the cotton candy we've all been eating
until I, at least, am sick to death
I'm furious with English-only laws
with Japanese-bashing celebrated
as some kind of special holy cleancut sport
Furious that anti-Semitism is as respectable as ever
& everybody who wants to talk about it must be a *pushy Jew*
I could kill those thousands of people who claim the nazi Holocaust
didn't happen
I'm angry that as these words rattle out of my mouth
I'm already cutting them back   cooling them off
taking the sting out because I'm afraid of what I might do
if I hear one more damn time
WHY are you so angry?
Raging that common sense & kindness are passé
not quite with it
Angry that breast cancer kills twice as many women
as men who have died of AIDS/SIDA but we're all
still paying attention to the poor men
as usual
I'm blowing my top about clear cuts, abuse of resources
abuse of workers, torture of animals for testing cosmetics

with the terrifying idea that wearing fur makes a woman sexy or special
with the largest slave labor force in the world which is called
the u.s. bureau of prisons
Sick of everyone watching light-filled shadows on a screen
more important than life
that your average citizen spends more time
adoring those shadows than speaking to their own children
I'm furious with my incoherence
my inability to affect almost everything in my life
I'm angry with everyone who's said some appallingly stupid thing
about peace pipes or pow wows or totem poles or tipis
Furious that the accepted ways to solve our pain
are to pay somebody to listen to us
or to adopt some party line without deviation
& preach it to everyone else
or to get high or to buy yet another piece of crap we don't really need
or to disappear into games
Angry with organized & disorganized religions which fill people's lives
with ignorant laws or hocus pocus or convince them that pain is holy
although I reserve most of my venom for the catholic church
which ruined my life with lies I'm still unraveling
I'm angry that none of us lives to our potential
that we've frightened into being the least we can be
to survive
Outraged that so much is swept under rugs
that we can barely walk
Furious that almost everyone still uses the world *blind*
to mean ignorant or insensitive or clumsy
that millions of trees are slaughtered to print romance novels or spy chillers
& every kind of wall street garbage
until I'm ashamed
to put words to paper at all
Most of us can hardly function
poisoned by corporate nonsense
assaulted with unnecessary chemicals
making somebody who hates us a nice fat profit
Angry that my back hurts all the time
from cleaning the houses of the lazy wealthy for 20 years
not one of whom is as intelligent, creative, or powerful as I am
Angry that I'm going to die this angry
& probably not be able to change a damn thing

Enraged that every place I go is inaccessible
even when they've altered the bathrooms inside because it's the law
when a chair still can't get up the outside stairs or in the door
At the braille signs inside elevators where there are none outside it
Furious with ignorance & apathy those smug cousins in every family
I can't shut my heart to the pain thudding all around us
Here in my hands are all the faces of those I've seen begging
in doorways, on freeway ramps, on sidewalks
begging for change for a meal or a drink
whose desperation is now against the law
This is just the scratched raw surface of my anger
which is fueled by the righteousness
of knowing we don't have to live this way
We could embrace our profound connections
and our deep differences
learn from each other
Honor each other
begin to live without torturing
If you aren't as angry as I am we probably shouldn't try
to talk to each other
because I'm furious with your fear of anger
I'm angry that others are always telling me
that they feel them same way I do but they're afraid to say so
or they don't know how
or they'd lose their job or their lover
If you can speak
you can be angry
if you can't speak bang your fork
If you're furious with me
because I haven't mentioned something
you're angry about
get busy & write it yourself
There is no such beast as too angry
I'm a canary down this mine of apathy
singing & singing my yellow throat on fire
with this sacred holy purifying
spirit of anger

For Ayofemi Faloyan

Man and Technics: A Contribution to a Philosophy of Life (1931)

Oswald Spengler

In [an] increasing interdependence lies the quiet and deep revenge of Nature upon the being that has wrested from her the privilege of creation. This petty creator *against* Nature, this revolutionary in the world of life, has become the slave of his creature. The Culture, the aggregate of artificial, personal, self-made life-forms, develops into a close-barred cage for these souls that would not be restrained. The beast of prey, who made others his domestic animals in order to exploit them, has taken himself captive. The great symbol of this fact is the human *house*.

That, and his increasing numbers, in which the individual disappears as unimportant. For it is one of the most fateful consequences of the human spirit of enterprise that the population multiplies. Where anciently the pack of a few hundred roamed, there is sitting a people of tens of thousands. There are scarcely any regions empty of men. People borders on people, and the mere *fact* of the frontier—the limits of one's own power—arouses the old instincts to hate, to attack, to annihilate. The frontier, of whatever kind it may be, even the intellectual frontier, is the mortal foe of the Will-to-Power.

It is not true that human technics saves labour. For it is an essential characteristic of the personal and modifiable technics of man, in contrast to genus-technics, that every discovery contains the possibility and *necessity* of new discoveries, every fulfilled wish awakens a thousand more, every triumph over Nature incites to yet others. The soul of this beast of prey is ever hungry, his will never satisfied—that the curse that lies upon this kind of life, but also the grandeur inherent in its destiny. It is precisely its best specimens that know least of quiet, happiness, or enjoyment. And no discoverer has ever accurately foreseen the *practical* effect of his act. The more fruitful the leader's work, the greater the need of executive hands. And so, instead of killing the prisoners taken from hostile tribes, men begin to enslave them, so as to exploit their bodily strength. This is the origin of Slavery, which must, therefore, be precisely as old as the slavery of domestic animals....

Man, evidently, was tired of merely having plants and animals and slaves to serve him, and robbing Nature's treasures of metal and stone, wood and yarn, of managing her water in canals and wells, of breaking her resistances with ships and roads, bridges and tunnels and dams. Now he meant, not merely to plunder her of her materials, *but to enslave and harness her very forces* so as to multiply his own strength. This monstrous and unparalleled idea is as old as the Faustian Culture itself. Already in the tenth century we meet with technical constructions of a wholly new sort. Already the steam engine, the steamship, and the air machine are in the thoughts of Roger Bacon and Albertus Magnus. And many a monk busied himself in his cell with the idea of *Perpetual Motion....*

Over a few decades most of the great forests have gone, to be turned into newsprint, and climatic changes have been thereby set afoot which imperil the land-economy of whole populations. Innumerable animal species have been extinguished, or nearly so, like the bison; whole races of humanity have been brought almost to vanishing-point, like the North American Indian and the Australian.

All things organic are dying in the grip of organization. An artificial world is permeating and poisoning the natural. The Civilization itself has become a machine that does, or tries to do, everything in mechanical fashion. We think only in horse-power now; we cannot look at a waterfall without mentally turning it into electric power; we cannot survey a countryside full of pasturing cattle without thinking of its exploitation as a source of meat-supply; we cannot look at the beautiful old handiwork of an unspoilt primitive people without wishing to replace it by a modern technical process. Our technical thinking must have its actualization, sensible or senseless. The luxury of the machine is the consequence of a necessity of thought. In last analysis, the machine is a *symbol*, like its secret ideal, perpetual motion—a spiritual and intellectual, but no vital necessity....

This machine-technics will end with the Faustian civilization and one day will lie in fragments, *forgotten*—our railways and steamships as dead as the Roman roads and the Chinese wall, our giant cities and skyscrapers in ruins like old Memphis and Babylon. The history of this technics is fast drawing to its inevitable close. It will be eaten up from within, like the grand forms of any and every Culture. When, and in what fashion, we know not.

In Search of Noble Ancestors (1992)

John Mohawk

Images of stone-age man that appear in commercial entertainment media tell us volumes about the ideas and ambitions of our time and culture. A not-bad example appearing in recent years was the movie starring Daryl Hannah called *Clan of the Cave Bear*. It offered a view of life of humans some 70 millennia ago. The character played by Hannah was allegedly a fully modern human—*Homo sapiens*. She was with a people who were not quite fully *Homo sapiens*. We are left to conclude they were Neanderthal, or neo-Neanderthal, or some such anthropological grouping.

The point is, they were human beings, existing seventy thousand years ago, as depicted by the combined minds of writer, director, producer, makeup artist and actor. The visuals on the screen were the fictional creations of fully modern man. Those images represent practically the only visions of earlier humans which many in this society are likely to see.

According to these visions, hunter-gatherer society was generally prelingual. The people spoke a protolanguage of guttural utterances and childlike gestures. If they had any complex thoughts, they were able to suppress expression of them through clumsy behavior. They appeared, quite frankly, to have suffered brain damage. Indeed, brain damage may have been afoot in these depictions, but it had nothing to do with life circa 70,000 B.C. The brain damage exists in our own time, and for good reason. Contemporary hierarchically socialized (wo)man feels obligated to present ancient peoples as brutes capable of only rudimentary intelligence, condemned to generally incoherent thinking. In addition, as depicted in this and really countless other films. early man is usually presented as a social undesirable, especially in relation to his ideas about male/female roles. Practically anyone who watches the vision unfold on the silver screen will be entranced by it, engulfed in the emotions it triggers. It is easy to forget it is a piece of fiction, wholly fabricated to entire and thrill and do whatever ancient man is capable of doing on film to separate the unwary viewer from the price of admission to the theater. I submit that exercises such as *Clan* are a reflection of current culture. Caveman-bashing is as old as the information that human culture preceded our culture. The old caricatures depicted Caveman carrying a club, wearing colorful animal skin garb, and dragging a woman by the hair into the cave. That image was a comment upon the people who drew it. It helped set the stage for later, more sophisticated efforts.

A phenomenon is at play here which illuminates the conservatism of culture and the disinformation techniques of class society which serve to form people's expectations and which limit their experience in the world. Contemporary society is filled with popular lies, masked sometimes in the form of art and sometimes in the form of serious study. A motion picture which made the rounds recently was *Native Son*, the story of a violent black man written in the 1940s by a black man. With apologies to the author, *Native Son* was exactly what white, middle-class chic intelligentsia wanted to read about the nature of a black man. It fit their mold, and they embraced it enthusiastically. I think that is why many Native American writers offer us stories about Indian alcoholism and hopelessness, and it is why such works win praise and prizes in the art world. I am not arguing here that there is not plenty of pain and oppression suffered by American Indians, but rather that there is a dark side to the critics. It is a dark side that enables people to embrace so emotionally *The Color Purple* because it reaffirms what they already believe to be true about "them"—in this case black men—to the point it is almost a cynically good economic bet for writers from a minority to write the most gruesome exposé of life in the bowels of their culture. That kind of writing allows art critics to point self-righteously to its power of emotion, its humanness. What is rarely represented in such novels/movies are so-called primitive people who have problem-solving skills, who are emotionally stable, who are likable—unless they are victims.

The primitive people of *Clan* are homely (one might say unkempt). They move awkwardly, uncertainly. They are hunters and gatherers who live miserably. The message in the visuals is strong: they are inferior. Hannah is the emerging future—better looking, somehow more "human." She violates a taboo of the tribe—she touches a weapon to save a child's life—and is punished. It's painful to watch.

It also is, in my humble opinion, a grossly unfair depiction of Paleolithic humankind. Permit me for a moment to indulge in a bit of awe directed at humans of the big-game hunting period. I don't want to sound as though the era was filled with Stone Age Conan the Barbarian types, for nothing could be further from the truth. However, the evidence does suggest that ancient hunters were excellent long-distance runners, had good skills at trapping and stampeding large game, and that they developed cultures which lasted an extremely long time as distinct material cultures. Modern humans owe a lot to that long period of hunter-gatherers, not the least of which was probably the ability to inhabit most of the areas of the earth. From the rain forests to the Arctic, the first humans to inhabit the globe were hunter-gatherers, and no one has successfully colonized any place that they did not live. Not bad, for folks whose (onscreen) vocabulary was limited to a few grunts.

To be perfectly honest, we have many gaps in our knowledge about ancient peoples. We don't know, for example, if they had complex languages. The stone artifacts, the bone fragments, the cave drawings are silent about tongues. It is possible they had a complex language or, more likely, hundreds of complex languages. It is possible that some words from their languages have survived into modern English. Anything is possible.

We don't know much about ancient people's religion either. That they had religion seems quite universally acknowledged. That it involved, at some point, the expectation of life or something akin to consciousness after death, we can postulate. We can believe that ancient people believed that animals possessed spirits as humans possessed spirits, and that the success of the hunt depended on placating some spirits in some manner.

What we can project about ancient humans, from our acquaintance with Stone Age people who survived into relatively modern times, is that they were dreamers, visionaries, beings who spent their time looking into the stars and asking questions. We know something else about ancient humans from modern primitives, such as Australian Aborigines: ancient people spent a lot of time thinking about their role in the nature of things.

What little we do know strongly suggests that Paleolithic humankind was resourceful, intelligent, courageous, athletic and artistic, among other virtues. Why, then, does there exist a powerful prejudice against our long-dead ancestors? Why are they defamed in much the same way contemporary people of other contemporary colors, religions and cultures are depicted in efforts to dehumanize, degrade and generally discredit them? Is Paleolithic (wo)man somehow a threat? Why are we presented with a vision of the common ancestor which offers that, 50 or so generations ago, all our ancestors were essentially dolts?

The answer, I suspect, lies in the way modern society wants to see itself. In twentieth century society, people are encouraged to look to the present as the Golden Age of Man, and the future as the Future Golden Age. We are socialized to believe that things are getting better. Never mind that all the indicators are that things are getting worse. Blot out from your mind the burgeoning populations of the Third World, the disappearance of uncontaminated groundwater in the United States, desertification worldwide, or the hole in the ozone layer. Try not to think of the fact that each of these resulted from some form of public policy undertaken and blessed by contemporary world leaders of contemporary world governments. The Establishment wants us to look forward now to completing the mission, and the vision of the mission is also found in novels and on film. Captain Kirk is no blithering idiot. Neither are the twenty-fourth-century Neanderthals, the Klingons. Things are getting better.

How few are the dissenting voices, how few have the strength to cry out from the crowd, "I don't believe it!" How few people seem to be able to look into the past and see something noble there. How few try.

I, for one, want to salute those ancients. I think it was pretty brave of them to challenge the largest elephant that ever lived, the woolly mammoth, with little more than a couple of sticks and a stone attached to one of them. And, more often than not, live to tell about it.

I think it was nobler to stalk and kill a thunderous beast than it is to order ground meat wrapped in paper and Styrofoam—meat that was grown on land needed by peasants who can't get enough protein to survive, irreplaceable rainforest land, which is being ground, literally, into hamburger for Burger King. I can understand why they gave thanks to the animal's spirit when they killed it, saw themselves as kin to the creatures they depended on, constructed reality around dreams and visions instead of computers and statistical lies. I can understand why their cultures lasted so long.

But I can't believe they were brutish and stupid. Those terms are relative, of course, but I don't think it's true. What we know about modern primitives is that they are absolutely sane and coherent people until an alien society abuses them beyond recognition.

Primitive people are subjected to bad press because they represent two ideas that are dangerous to the powers that be. First, they are in the past and the powers that be act as though it is always and irreversibly harmful to think positively about the ancient past. I suggest exactly the opposite. Given the choice of finding our species' identity in clear-thinking, caring, sharing, and fully human ancestors, as opposed to creatures who act like members of an outlaw motorcycle gang in animal skins, I think there is little question which is healthier.

The vision of crude ancestors whose crudeness penetrates far into the era of fully modern human beings gives us permission to be less than we are capable of being. Thus dismissed, we are discouraged from thinking about their lives, their thoughts, their aspirations for their progeny (ourselves), our connections to them. Had we considered the ancients to be intelligent, resourceful, thoughtful, and in every positive way human, we would also be encouraged to respect their thoughts, their lives, their evolution into our time. We would be empowered to think as they thought, to share the responsibilities they envisioned, and to share their dreams.

Novels like *Lord of the Flies*—required reading in my adolescence—present quite a different message about the nobility of the human species. Movies like *Thunderdome* carry that message into a desolate and recycled-primitive future. Interestingly, danger and crisis often bring out the best in people, not the worst. The idea that the past was brutish and the future without civilization is brutish serves to reinforce the myth that humanity is less human without the rules and guidance of modern hierarchical society.

The second message, the one I think is even more destructive than the first, is that all primitives do is essentially trite, built on useless superstition, and unworkable. Primitive people paid a lot of attention to dreams, to dreaming, to thinking about the universe, the interrelationship of humans and animals, to the transformation of the consciousness of animals to human. Primitives were a complex lot. Within that complexity lies the potential for a whole realm of consciousness which modern society finds unacceptable, indeed dangerous.

The reason for this, of course, is that the glue of modern society depends, to a considerable extent, on the premise that people are not required to do their own thinking. If primitive people thought about everything, spent time thinking about relationship of wolf spirits to water spirits, it is also a fair generalization that modern people think about practically nothing. The average individual has been socialized to believe that all the thinking necessary to the great mysteries of life is being done for him or her.

That was probably the first crime done by the advent of modern civilization. Civilizations created specialists, castes, and class systems, which socialized people to the belief that all the thinking about important subjects, from God to hydrology, was being taken care of by specialists and that the individual had no obligation to spend any time thinking of these things. The individual's lot in civilization is to pay taxes and leave thinking to those in the know.

One has to believe this is why civilizations are destroying the world. Humans inhabited the great forests of the northern hemisphere for thousands of years, but only two centuries of European expansion into North America and the forest is, for all practical purposes, destroyed. One is tempted to blame the marketplace for this destruction, but a destruction so complete cannot even be explained by the market. Every last passenger pigeon, of a population of birds that once blackened the skies, was killed by a people who do not find in themselves the necessity to think about the relationship of pigeons to man, the future, the land.

Civilized people are socialized to leave thinking to others; indeed, when Westerners come into contact with "primitive" peoples, they view with astonishment the practices that require every individual to spend time thinking about a vast array of plants, animals, waters, even volcanoes. "Superstition!" they shout, not able to recognize a culture that socializes people to think about what they do.

Primitive people knew they had a responsibility to think, to feel, to dream. Had they not evolved a culture which promoted and rewarded thinking as a human activity, our species would not have evolved as it did. Culture, after all, is learned behavior. Civilization has demanded that the individual cease thinking about what he does.

SECTION V

THE RESISTANCE TO
CIVILIZATION

INDUSTRIAL VALLEY

If Man becomes an animal again, his arts, his loves, and his play must also become purely "natural" again. Hence it would have to be admitted that after the end of History, men would construct their edifices and works of art as birds build their nests and spiders spin their webs, would perform musical concerts after the fashion of frogs and cicadas, would play like young animals, and would indulge in love like adult beasts.

—Alexandre Kojève (1946)

We move now to offerings that attempt to light the way beyond civilization, to sources and modes of resistance and renewal. The deconstructionist Derrida applies the tactic of placing literary elements "under erasure"; here are some considerations for doing so to civilization.

Julia Kristeva recently rejected the postmodern refusal of narrative, or refusal of viewing the totality, in this way:

> Psychoanalysis goes against the grain of the modern convenience that calls attention not to the end of the Story of Civilization, but to the end of the possibility of telling a story. Nevertheless, this end and this convenience are beginning to overwhelm us, and we have been led to criticize and reject them.

One need not adopt psychoanalysis as the answer to the postmodern dead end, but Kristeva's conclusion is most important regardless, in refusing to accept an end to possibilities.

Another necessary rejection of limits concerns a more general or typical defeatism, in parallel to that of postmodernism. From a recent work by former 1960s activist Gregory Calvert:

> It is, I believe, an error (and the weakness of certain kinds of anarchist utopianism) to assume that humanity can somehow return to the "organic" or "natural" societies of the neolithic world, or that there is an end to politics. Human beings have left forever their neolithic past and life in the human realm can never be a simple return to nature.

He means, of course, paleolithic not neolithic, for the latter is synonymous with the arrival of civilization. If there is consensus among authors represented in this section, however, it lies in their rejection of the argument that a "return to nature" is impossible. Calvert's caveat is just another way of saying, "Here is civilization: accept it."

The summer 1995 issue of the British Marxist journal *Aufheben* acknowledges that

> civilization is under attack. A new critical current has emerged in recent years, united by an antagonism towards all tendencies that seem to include "progress" as part of their programme.

Indeed, a question heard with increasing frequency asks how much more progress we and the planet must endure. This critique challenges the basic categories and dynamics of civilization, and demands an altogether different present and future.

Avoiding Social and Ecological Disaster: The Politics of World Transformation (1994)

Rudolf Bahro

What is exterminism?

In order to furnish a basis for resistance to rearmament plans, the visionary British historian E.P. Thompson wrote an essay in 1980 about *exterminism*, as the last stage of civilization. Exterminism doesn't just refer to military overkill, or to the neutron bomb—it refers to *industrial civilization as a whole*, and to many aspects of it, not just the material ones—although these are the first to be noticed. It made sense that the *ecopax* movement in Germany began not with nuclear weapons, but with nuclear power stations, and seemingly even less harmful things. Behind the various resistance movements stood the unspoken recognition that in the set of rules guiding the evolution of our species, death has made its home.

Thompson's statements about the "increasing determination of the extermination process," about the "last dysfunction of humanity, its total self-destruction," characterize the situation as a whole. The number of people who are damned and reduced to misery has increased unbelievably with the spread of industrial civilization. Never in the whole of history have so many been sacrificed to hunger, sickness, and premature death as is the case today. It is not only their number which is growing, but also their proportion of the whole of humanity. As an inseparable consequence of military and economic progress we are in the act of destroying the biosphere which gave birth to us.

To express the exterminism thesis in Marxian terms, one could say that the relationship between productive and destructive forces is turned upside down. Like others who looked at civilization as a whole, Marx had seen the trail of blood running through it, and that "civilization leaves deserts behind it." In ancient Mesopotamia it took 1,500 years for the land to grow salty, and this was only noticed at a very late stage, because the process was so slow. Ever since we began carrying on a productive material exchange with nature, there has been this destructive side. And today we are forced to think apocalyptically, not because of culture-pessimism, but because this destructive side is gaining the upper hand.

I would like straightaway to emphasize that the problem ultimately does not lie in the perversions and associated monstrosities of Auschwitz and Hiroshima, in neurotic lust for destruction or for human or animal torture. It lies in quantitative success, and in the direction that our civilization took in its heyday. This success is not at all unlike that of a swarm of locusts. Our higher level of consciousness has furthered development, but has had no part in determining scale or goal. In general the logic of self-extermination works blindly, and its tools are not the ultimate cause.

For centuries the problem has remained below the threshold of consciousness for the vast majority of people. In the *Communist Manifesto*, Marx and Engels evaluate the capitalistic preparatory work for the desired classless society:

Through the exploitation of the world market, the bourgeoisie has given a cosmopolitan pattern to the production of all countries. To the great dismay of reactionaries it has pulled the national basis out from under the feet of industry. The age-old national industries have been annihilated and continue to be annihilated daily. They get pushed aside by new industries, the introduction of which becomes a life-issue for all civilized nations.

These new industries don't make use of domestic raw materials, but process raw materials from the remotest regions, and their products are used not only in the land of their production, but equally in all parts of the world. In the place of needs which can be satisfied by domestic production come new ones, which demand for their satisfaction the products of the remotest lands and climates. In the place of the old national and local self-sufficiency and isolation comes traffic in all directions, and a dependence on all sides of nations upon each other....

By means of enormously increased ease of communication the bourgeoisie draws all nations, even the most barbaric, into civilization. The cheap prices of its wares are the heavy artillery by which all Chinese walls are shot down, by which the most stubborn barbarian xenophobia must capitulate.

As we see today, this is written on account of civilized worker-interests, and is a clearly 'social-imperialistic' text. The concern is with the proletarian take-over of business in this civilization, and social democracy, or even more the trade unions, and the legitimate heirs of this programme, to whose basic cultural themes they adhere unbrokenly.

Wolfram Ziegler has developed a scale which measures, with brilliant simplicity, the *total load* we are placing on the biosphere, in order to bring about the 'good life' or 'standard of living', and on this basis to defend the 'social peace' of the rich Metropolis, which is certainly being ever more strongly threatened by ecological panic. Ziegler's starting point is that the decisive lever in our attack on nature is the use of technically-prepared imported energy. The poisoning and destruction of nature is bound up with this material throughout, with the putting to work of our energy-slaves.

For this reason Ziegler takes the amount of energy used per square kilometer per day and multiplies it by a 'damage-equivalent' for the amount of matter-transformation, and impact on nature, in each region. In this way he arrives at a figure for the load on the biosphere measured in equivalent kilowatt-hours per square kilometer per day. This figure is far in excess of the raw energy use because the toxic and noxious effects are factored in. Today in Germany we are impacting the environment to the extent of 40,000 KWh/km²/day [103,600 KWh/sq. mile/day] with real energy use alone—that is, without reckoning in the damage factor. This is about ten times as much as it was a hundred years ago.

Exactly a hundred years ago, the rate of dying out of biological species began to increase exponentially; as a result of which in the mid-1980s a species vanished every day, and by the year 2000 this will have increased to *a species every hour*. We are monopolizing the earth for our species alone. We began this with the geographical surface, which we don't only reduce in area, but divide up to such an extent that ecotopes lose their wholeness, and the critical number of individuals of any species is reached, such that they cannot share the same living space.

Ziegler has calculated that in Germany the total weight of our bodies averages out at 150 kg per hectare [134 lbs per acre], while all other animals including birds weigh only 8–8.5 kg per hectare [7–7.5 lbs per acre]. This excludes the domesticated animals we exploit, which account for a further 300 kg per hectare [267 lbs per acre]—however they don't belong to *themselves*, but to us. In addition to this, we have at least a further 2,000 kg per hectare [1,780 lbs per acre] of technical structures for our transport systems alone, and the lion's share of this is taken by the automobile.

Even though we no longer feel any natural solidarity with the rest of life, we nevertheless depend, for our biological existence, upon the species variety of plants and animals. Our 'anthropogenic' technical monocultures of 'useful' plants and animals are perhaps the most persistent instruments of suicide that we use. The dying off of species is the most fundamental indicator of the general exterminating tendency: the overgrowth of the industrial system has pushed it to a galloping rate.

For Ziegler a load of about 4,000 real KWh/km²/day [10,360 KWh/sq mile/day] is the threshold at which we finally left ecological stability behind us. It is about where we were 100 years ago, before the rate of dying off of species began to increase, and before—a quarter of a century later—the first organizations for the protection of nature began to react.

It is thus no longer a question of *democratic decision*, but of *natural necessity*, that we should reduce harmful end-products of energy and materials-consumption by a factor of *ten*. By more intuitive methods I had reached the same estimate of the order of magnitude of the necessary reduction, by reflecting on what would happen if the whole of humanity were to lay claim to our level of packaged living.

Environmental protection is a 'solution' one would expect from the priesthood (this time a scientific one) of a declining culture: one more floor on the deficit-structure, which would only increase its load. Ziegler demonstrates compellingly that this is not to be done with technical environmental protection alone, because the energy and material demands of such measures would detract from the load reductions they would achieve, and ultimately cancel them entirely.

Thus in the final analysis environmental protection as a supplementary strategy is only a further stimulus to the economic arms race, whereby the mass of the Megamachine is made to grow, both on the investment and the consumption side. Janicke has demonstrated this from the point of view of costs, basing his work on Kapp (1972). Environmental protection procures a last 'Green' legitimacy for the industrial system, for a short while.

While we protect the environment at critical points, the whole front of stress on the natural order moves unflinchingly ahead. A hundred environment-protecting motors each having only 66% of the damaging effect of earlier models still do more damage than 50 of the earlier models.

Messages about the environment-friendliness of industry, seen today on TV screens and in magazines, create a fatally false impression. For example, via foodstuffs alone we come into contact with about 10,000 chemicals, and in daily life with about 100,000 of them, in industrial nations. Propaganda deceptively plays down this synthetic aspect of civilized life. We can adapt to plastics thanks to the much-praised plasticity of human nature, which we also have to thank for civilization! By exercising our drive to activity, our passion for work, we altogether ruin our entire original fund of resources. In this context ecological market economy is only a new addition to the logic of self-extermination. Its immediate effect is to lower the level of product-or-technology-specific environmental damage, but the long-term overall effect is to *increase* it.

FUTURE PRIMITIVE (1994)

JOHN ZERZAN

ivision of labor, which has had so much to do with bringing us to the present global crisis, works daily to prevent our understanding the origins of this horrendous present. Mary Lecron Foster (1990) surely errs on the side of understatement in allowing that anthropology is today "in danger of serious and damaging fragmentation." Shanks and Tilley (1987b) voice a rare, related challenge: "The point of archaeology is not merely to interpret the past but to change the manner in which the past is interpreted in the service of social reconstruction in the present." Of course, the social sciences themselves work against the breadth and depth of vision necessary to such a reconstruction. In terms of human origins and development, the array of splintered fields and sub-fields—anthropology, archaeology, paleontology, ethnology, paleobotany, ethnoanthropology, etc., etc.—mirrors the narrowing, crippling effect that civilization has embodied from its very beginning.

Nonetheless, the literature can provide highly useful assistance, if approached with an appropriate method and awareness and the desire to proceed past its limitations. In fact, the weakness of more or less orthodox modes of thinking can and does yield to the demands of an increasingly dissatisfied society. Unhappiness with contemporary life becomes distrust with the official lies that are told to legitimate that life, and a truer picture of human development emerges. Renunciation and subjugation in modern life have long been explained as necessary concomitants of "human nature." After all, our pre-civilized existence of deprivation, brutality, and ignorance made authority a benevolent gift that rescued us from savagery. "Cave man" and "Neanderthal" are still invoked to remind us where we would be without religion, government and toil.

This ideological view of our past has been radically overturned in recent decades, through the work of academics like Richard Lee and Marshall Sahlins. A nearly complete reversal in anthropological orthodoxy has come about, with important implications. Now we can see that life before domestication/agriculture was in fact largely one of leisure, intimacy with nature, sensual wisdom, sexual equality, and health. This was our human nature, for a couple of million years, prior to enslavement by priests, kings, and bosses....

To "define" a disalienated world would be impossible and even undesirable, but I think we can and should try to reveal the unworld of today and how it got this way. We have taken a monstrously wrong turn with symbolic culture and division of labor, from a place of enchantment, understanding and wholeness to the absence we find at the heart of the doctrine of progress. Empty and emptying, the logic of domestication, with its demand to control everything, now shows us the ruin of the civilization that ruins the rest. Assuming the inferiority of nature enables the domination of cultural systems that soon will make the very earth uninhabitable.

Postmodernism says to us that a society without power relations can only be an abstraction (Foucault, 1982). This is a lie unless we accept the death of nature and renounce what once was and what we can find again. Turnbull spoke of the intimacy between Mbuti people and the forest, dancing almost as if making love to the forest. In the bosom of a life of equals that is no abstraction, that struggles to endure, they were "dancing with the forest, dancing with the moon."

News from Nowhere (1995)

William Morris

We went up a paved path between the roses, and straight into a very pretty room, panelled and carved, and as clean as a new pin; but the chief ornament of which was a young woman, light-haired and grey-eyed, but with her face and hands and bare feet tanned quite brown with the sun. Though she was very lightly clad, that was clearly from choice, not from poverty, though these were the first cottage-dwellers I had come across; for her gown was of silk, and on her wrists were bracelets that seemed to me of great value. She was lying on a sheep-skin near the window, but jumped up as soon as we entered, and when she saw the guests behind the old man, she clapped her hands and cried out with pleasure, and when she got us into the middle of the room, fairly danced round us in delight of our company.

"What!" said the old man, "you are pleased, are you, Ellen?"

The girl danced up to him and threw her arms round him, and said: "Yes I am, and so ought you to be, grandfather."

"Well, well, I am," said he, "as much as I can be pleased. Guests, please be seated."

This seemed rather strange to us; stranger, I suspect, to my friends than to me; but Dick took the opportunity of both the host and his grand-daughter being out of the room to say to me, softly: "A grumbler: there are a few of them still. Once upon a time, I am told, they were quite a nuisance."

The old man came in as he spoke and sat down beside us with a sigh, which, indeed, seemed fetched up as if he wanted us to take notice of it; but just then the girl came in with the victuals, and the carle missed his mark, what between our hunger generally and that I was pretty busy watching the grand-daughter moving about as beautiful as a picture.

Everything to eat and drink, though it was somewhat different to what we had had in London, was better than good, but the old man eyed rather sulkily the chief dish on the table, on which lay a leash of fine perch, and said:

"H'm, perch! I am sorry we can't do better for you, guests. The time was when we might have had a good piece of salmon up from London for you; but the times have grown mean and petty."

"Yes, but you might have had it now," said the girl, giggling, "if you had known that they were coming."

"It's our fault for not bringing it with us, neighbors," said Dick, good-humoredly. "But if the times have grown petty, at any rate the perch haven't; that fellow in the middle there must have weighed a good two pounds when he was showing his dark stripes and red fins to the minnows yonder. And as to the salmon, why, neighbor, my friend here, who comes from the outlands, was quite surprised yesterday morning when I told him we had plenty of salmon at Hammersmith. I am sure I have heard nothing of the times worsening."

He looked a little uncomfortable. And the old man, turning to me, said very courteously:

"Well, sir, I am happy to see a man from over the water; but I really must appeal to you to say whether on the whole you are not better off in your country; where I suppose, from what our guest says, you are brisker and more alive, because you have not wholly got rid of competition. You see, I have read not a few books of the past days, and certainly *they* are much more alive than those which are written now; and good sound unlimited competition was the condition under which they were written,—if we didn't know that from the record of history, we should know it from the books themselves. There is a spirit of adventure in them, and signs of a capacity to extract good out of evil which our literature quite lacks now; and I cannot help thinking that our moralists and historians exaggerate hugely the unhappiness of the past days, in which such splendid works of imagination and intellect were produced."

Clara listened to him with restless eyes, as if she were excited and pleased; Dick knitted his brow and looked still more uncomfortable, but said nothing. Indeed, the old man gradually, as he warmed to his subject, dropped his sneering manner, and both spoke and looked very seriously.

But the girl broke out before I could deliver myself of the answer I was framing:

"Books, books! always books, grandfather! When will you understand that after all it is the world we live in which interests us; the world of which we are a part, and which we can never love too much? Look!" she said, throwing open the casement wider and showing us the white light sparkling between the black shadows of the moonlit garden, through which ran a little shiver of the summer night-wind, "look! these are our books in these days!—and these," she said, stepping lightly up to the two lovers and laying a hand on each of their shoulders; "and the guest there, with his oversea knowledge and experience; yes, and even you, grandfather" (a smile ran over her face as she spoke), "with all your grumbling and wishing yourself back again in the good old days,—in which, as far as I can make out, a harmless and lazy old man like you would either have pretty nearly starved, or have had to pay soldiers and people to take the folks' victuals and clothes and houses away from them by force. Yes, these are our books; and if we want more, can we not find work to do in the beautiful buildings that we raise up all over the country (and I know there was nothing like them in past times), wherein a man can put forth whatever is in him, and make his hands set forth his mind and his soul."

She paused a little, and I for my part could not help staring at her, and thinking that if she were a book, the pictures in it were most lovely. The color mantled in her delicate sunburnt cheeks; her grey eyes, light amidst the tan of her face, kindly looked on us all as she spoke. She paused, and said again:

"As for your books, they were well enough for times when intelligent people had but little else in which they could take pleasure, and when they must needs supplement the sordid miseries of their own lives with imaginations of the lives of other people. But I say flatly that in spite of all their cleverness and vigor, and capacity for story-telling, there is something loathsome about them. Some of them, indeed, do here and there show some feeling for those whom the history-books call 'poor,' and of the misery of whose lives we have some inkling; but presently they give it up, and towards the end of the story we must be contented to see the hero and heroine living happily in an island of bliss on other people's troubles; and that after a long series of sham troubles (or mostly sham) of their own making, illustrated by dreary introspective nonsense about their feelings and aspirations, and all the rest of it; while the world must even then have gone on its way, and dug and sewed and baked and built and carpentered round about these useless—animals."

"There!" said the old man, reverting to his dry sulky manner again. "There's eloquence! I suppose you like it?"

"Yes," said I, very emphatically.

"Well," said he, "now the storm of eloquence has lulled for a little, suppose you answer my question?—that is, if you like, you know," quoth he, with a sudden access of courtesy.

"What question?" said I. For I must confess that Ellen's strange and almost wild beauty had put it out of my head.

Said he: "First of all (excuse my catechizing), is there competition in life, after the old kind, in the country whence you come?"

"Yes," said I, "it is the rule there." And I wondered as I spoke what fresh complications I should get into as a result of this answer.

"Question two," said the carle: "Are you not on the whole much freer, more energetic—in a word, healthier and happier—for it?"

I smiled. "You wouldn't talk so if you had any idea of our life. To me you seem here as if you were living in heaven compared with us of the country from which I came."

"Heaven?" said he: "you like heaven, do you?"

"Yes," said I—snappishly, I am afraid; for I was beginning rather to resent his formula.

"Well, I am far from sure that I do," quoth he. "I think one may do more with one's life than sitting on a damp cloud and singing hymns."

I was rather nettled by this inconsequence, and said: "Well, neighbor, to be short, and without using metaphors, in the land whence I come, where the competition which produced those literary works which you admire so much is still the rule, most people are thoroughly unhappy; here, to me at least, most people seem thoroughly happy."

"No offense, guest—no offense," said he; "but let me ask you; you like that, do you?"

His formula, put with such obstinate persistence, made us all laugh heartily; and even the old man joined in the laughter on the sly. However, he was by no means beaten, and said presently:

"From all I can hear, I should judge that a young woman so beautiful as my dear Ellen yonder would have been a lady, as they called it in the old time, and wouldn't have had to wear a few rags of silk as she does now, or to have browned herself in the sun as she has to do now. What do you say to that, eh?"

Here Clara, who had been pretty much silent hitherto, struck in and said: "Well, really, I don't think that you would have mended matters, or that they want mending. Don't you see that she is dressed deliciously for this beautiful weather? And as for the sun-burning of your hay-fields, why, I hope to pick up some of that for myself when we get a little higher up the river. Look if I don't need a little sun on my pasty white skin!"

And she stripped up the sleeve from her arm and laid it beside Ellen's who was now sitting next her. To say the truth, it was rather amusing to me to see Clara putting herself forward as a town-bred fine lady, for she was as well-knit and clean-skinned a girl as might be met with anywhere at the best. Dick stroked the beautiful arm rather shyly, and pulled down the sleeve again, while she blushed at his touch; and the old man said laughingly: "Well, I suppose you *do* like that; don't you?"

Ellen kissed her new friend, and we all sat silent for a little, till she broke out into a sweet shrill song, and held us all entranced with the wonder of her clear voice; and the old grumbler sat looking at her lovingly. The other young people sang also in due time; and then Ellen showed us to our beds in small cottage chambers, fragrant and clean as the ideal of the old pastoral poets; and the pleasure of the evening quite extinguished my fear of the last night, that I should wake up in the old miserable world of worn-out pleasures, and hopes that were half fears....

All along, though those friends were so real to me, I had been feeling as if I had no business amongst them: as though the time would come when they would reject me, and say, as Ellen's last mournful look seemed to say, "No, it will not do; you cannot be of us; you belong so entirely to the unhappiness of the past that our happiness even would weary you. Go back again, now you have seen us, and your outward eyes have learned that in spite of all the infallible maxims of your day there is yet a time of rest in store for the world, when mastery has changed into fellowship—but not before. Go back again, then, and while you live you will see all round you people engaged in making others live lives which are not their own, while they themselves care nothing for their own real lives—men who hate life though they fear death. Go back and be the happier for having seen us, for having added a little hope to your struggle. Go on living while you may, striving, with whatsoever pain and labor needs must be, to build up little by little the new day of fellowship, and rest, and happiness."

Yes, surely! and if others can see it as I have seen it, then it may be called a vision rather than a dream.

"Feral Revolution"

Feral Faun

When I was a very young child, my life was filled with intense pleasure and a vital energy that caused me to feel what I experienced to the full. I was the center of this marvelous, playful existence and felt no need to rely on anything but my own living experience to fulfill me.

I felt intensely, I experienced intensely, my life was a festival of passion and pleasure. My disappointments and sorrows were also intense. I was born a free, wild being in the midst of a society based upon domestication. There was no way that I could escape being domesticated myself. Civilization will not tolerate what is wild in its midst. But I never forgot the intensity that life could be. I never forgot the vital energy that had surged through me. My existence since I first began to notice that this vitality was being drained away has been a warfare between the needs of civilized survival and the need to break loose and experience the full intensity of life unbound.

I want to experience this vital energy again. I want to know the free-spirited wildness of my unrepressed desires realizing themselves in festive play. I want to smash down every wall that stands between me and the intense, passionate life of untamed freedom that I want. The sum of these walls is everything we call civilization, everything that comes between us and the direct, participatory experience of the wild world. Around us has grown a web of domination, a web of mediation that limits our experience, defining the boundaries of acceptable production and consumption.

Domesticating authority takes many forms, some of which are difficult to recognize. Government, capital and religion are some of the more obvious faces of authority. But technology, work, language with its conceptual limits, the ingrained habits of etiquette and propriety—these too are domesticating authorities which transform us from wild, playful, unruly animals into tamed, bored, unhappy producers and consumers. These things work in us insidiously, limiting our imaginations, usurping our desires, suppressing our lived experience. And it is the world created by these authorities, the civilized world, in which we live. If my dream of a life filled with intense pleasure and wild adventure is to be realized, the world must be radically transformed, civilization must fall before expanding wilderness, authority must fall before the energy of our wild freedom. There must be—for want of a better word—a revolution.

But a revolution that can break down civilization and restore the vital energy of untamed desire cannot be like any revolution of the past. All revolutions to date have centered around power, its use and redistribution. They have not sought to eradicate the social institutions that domesticate; at best they have only sought to eradicate the power relationships within those institutions. So revolutionaries of the past have aimed their attacks at the centers of power seeking to overthrow it.

Focused on power, they were blind to the insidious forces of domination that encompass our daily existence—and so, when successful at overthrowing the powers that be, they ended up re-creating them. To avoid this, we need to focus not on power, but on our desire to go wild, to experience life to the full, to know intense pleasure and wild adventure. As we attempt to realize this desire, we confront the real forces of domination, the forces that we face every moment of every day. These forces have no single center that can be overthrown. They are a web that binds us. So rather than trying to overthrow the powers that be, we want to undermine domination as we confront it every day, helping the already collapsing civilization to break down more quickly—and as it falls, the centers of power will fall with it. Previous revolutionaries have only explored the well-mapped territories of power. I want to explore and adventure in the unmapped, and unmappable, territories of wild freedom. The revolution that can create the world I want has to be a feral revolution.

There can be no programs or organizations for feral revolution, because wildness cannot spring from a program or organization. Wildness springs from the freeing of our instincts and desires, from the spontaneous expression of our passions. Each of us has experienced the processes of domestication, and this experience can give us the knowledge we need to undermine civilization and transform our lives. Our distrust of our own experience is probably what keeps us from rebelling as freely and actively as we'd like. We're afraid of fucking up, we're afraid of our own ignorance. But this distrust and fear have been instilled in us by authority. It keeps us from really growing and learning. It makes us easy targets for any authority that is ready to fill us. To set up "revolutionary" programs is to play on this fear and distrust, to reinforce the need to be told what to do. No attempt to go feral can be successful when based on such programs. We need to learn to trust and act upon our own feelings and experiences, if we are ever to be free.

So I offer no programs. What I will share is some thoughts on ways to explore. Since we all have been domesticated, part of the revolutionary process is a process of personal transformation. We have been conditioned not to trust ourselves, not to feel completely, not to experience life intensely. We have been conditioned to accept the humiliation of work and pay as inescapable, to relate to things as resources to be used, to feel the need to prove ourselves by producing. We have been conditioned to expect disappointment, to see it as normal, not to question it. We have been conditioned to accept the tedium of civilized survival rather than breaking free and really living. We need to explore ways of breaking down this conditioning, of getting as free of our domestication as we can now. Let's try to get so free of this conditioning that it ceases to control us and becomes nothing more than a role we use when necessary for survival in the midst of civilization as we strive to undermine it.

In a very general way, we know what we want. We want to live as wild, free beings in a world of wild, free beings. The humiliation of having to follow rules, of having to sell our lives away to buy survival, of seeing our usurped desires transformed into abstractions and images in order to sell us commodities fills us with rage. How long will we put up with this misery? We want to make this world into a place where our desires can be immediately realized, not just sporadically, but normally. We want to re-eroticize our lives. We want to live not in a dead world of resources, but in a living world of free wild lovers. We need to start exploring the extent to which we are capable of living these dreams in the present without isolating ourselves. This will give us a clearer understanding of the domination of civilization over our lives, an understanding which will allow us to fight domestication more intensely and so expand the extent to which we can live wildly.

Attempting to live as wildly as possible now will also help break down our social conditioning. This will spark a wild prank-ishness in us which will take aim at all that would tame it, undermining civilization and creating new ways of living and sharing with each other. These explorations will expose the limits of civilization's domination and will show its inherent opposition to freedom. We will discover possibilities we have never before imagined—vast expanses of wild freedom. Projects, ranging from sabotage and pranks that expose or undermine the dominant soci-ety, to the expansion of wilderness, to festivals and orgies and general free sharing, can point to amazing possibilities.

Feral revolution is an adventure. It is the daring exploration of going wild. It takes us into unknown territories for which no maps exist. We can only come to know these territories if we dare to explore them actively. We must dare to destroy whatever destroys our wildness and to act on our instincts and desires. We must dare to trust in ourselves, our experiences and our passions. Then we will not let ourselves be chained or penned in. We will not allow ourselves to be tamed. Our feral energy will rip civilization to shreds and create a life of wild freedom and intense pleasure.

"Don't Eat Your Revolution! Make It!" (1995)

O n November 7 all the Stalinists gathered on Oktyabrskaya Square in Moscow under the still-standing huge Lenin. Later they headed to Red Square, followed by some small groups of Trotskyists and even some anarchists who successfully created an illusion that "young people" were with them. While the latters' red-n-black banners and dull papers were surely unable to destroy the traditional way of celebrating the October revolution day, the assault came from where nobody expected it. As the demonstrators were going down their route they inevitably confronted a small group of some vocal revolutionaries who were standing on the side of the road under the banner of "primitive communism" (made of fake red fur) who were fearlessly banging their big drum. As the crowd passed by them and tried to find out what they wanted to say, the atmosphere was getting more and more tense. "The Blind" (that is the name of an artistic group) announced that people who gathered at the demo had nothing to do with revolution, that they "ate" their revolution and shitted it out, that the only thing they can do is ask the government for more money. Their poster announced "Proletarians of all lands, enough eating!" Instead of asking the government for more money, The Blind announced, we should make another revolution and realize the real—primitive—communism. We should care less about material things and instead turn our attention to people around us. After the demo passed and the shouts "They are Jews, Zionists! Go back to Israel!" addressed to The Blind finally dissolved in the air, the "primitive communists" headed back home and were stopped by police, who asked for their IDs. In spite of that, the affair ended peacefully.

"THE MACHINE IN OUR HEADS" (1997)

GLENN PARTON

Introduction

The environmental crisis consists of the deterioration and outright destruction of micro and macro ecosystems worldwide, entailing the elimination of countless numbers of wild creatures from the air, land, and sea, with many species being pushed to the brink of extinction, and into extinction. People who passively allow this to happen, not to mention those who actively promote it for economic or other reasons, are already a good distance down the road to insanity. Most people do not see, understand, or care very much about this catastrophe of the planet because they are overwhelmingly preoccupied with grave psychological problems. The environmental crisis is rooted in the psychological crisis of the modern individual. This makes the search for an eco-psychology crucial; we must understand better what terrible thing is happening to the modern human mind, why it is happening, and what can be done about it.

Deep Thinking

The solution to the global environmental crisis we face today depends far less on the dissemination of new information than it does on the re-emergence into consciousness of old ideas. Primitive ideas or tribal ideas—kinship, solidarity, community, direct democracy, diversity, harmony with nature—provide the framework or foundation of any rational or sane society. Today, these primal ideas, gifts of our ancestral heritage, are blocked from entering consciousness. The vast majority of modern people cannot see the basic truths that our ancient ancestors knew and that we must know again, about living within the balance of nature. We are lost in endless political debates, scientific research, and compromises because what is self-evident to the primitive mind has been forgotten.

For hundreds of thousands of years, until the beginning of civilization about 10,000 years ago, humans lived in tribal societies, which produced tribal consciousness—a set of workable ideas or guiding principles concerning living together successfully on a diverse and healthy planet. The invasion of civilization into one tribal locale after another, around the globe, has been so swift and deadly that we may speak of the trauma of civilization. Because tribal peoples were unprepared and unable to deal with the onslaught of civilization, tribal consciousness was driven underground, becoming something forbidden and dangerous. Conquered peoples became afraid to think and act according to the old ways, on pain of death. There is much fear that lies at the origin of civilization.

Ontogeny recapitulates phylogeny—that is to say, the development of the individual is an abbreviated repetition of the development of the species. In childhood, a modern person travels an enormous distance between stone-age primitive creature and responsible contemporary citizen. When confronted with the awesome power of civilization whose first representatives are parents, teachers, priests (and, later on, police officers, legislators and bosses)—the child faces, psychologically, the same situation as its tribal ancestors—namely, conform to the dictates of civilization or die. The helplessness of childhood makes the threat of bodily harm or loss of love, which is used by the parents and others to enforce civilized morality and civilized education, a traumatic experience. The developing little person becomes afraid to express its own tribal nature. There is much fear that lies at the bottom of becoming a civilized adult.

When the child becomes aware of ideas and impulses that oppose the dictates of civilization, s/he experiences anxiety, which is the signal for danger. It is not the insights and urges themselves that the child fears, but rather the reaction to them on the part of those in charge. Since the child cannot escape from those who control its life, s/he runs away from dangerous thoughts and feelings. In other words, the child institutes repression of its primitive self. Tribal ideas are now isolated, cut off from awareness, and unable to properly influence the future course of events.

The trauma or inescapable terror of civilization is responsible for the derangement of reason. That inner dialogue in the human mind that is the hallmark of self-consciousness has ceased, because the depth-dimension of reflective thought, which is the primitive mind, has been silenced. Modern people no longer hear their own primal voice, and without interaction between new ideas and old ideas, the demands of the individual and the demands of the tribe (and species), there is no deep thinking. On the contrary, when reason is cut off at the roots, it becomes shallow, unable to determine what is of true value in life.

The passage of tribal ideas from the oldest and deepest layer of the mind into individual consciousness is part of the natural, normal functioning of the human mind. Deep thinking is not the result of education; it is innate, our birthright as *Homo sapiens*. What civilization has done is to disrupt the free flow of ideas in the human mind by shutting down the primitive mind through traumatic socialization. In such a situation, cut off from the time-tested and proven ideas of prehistory, reason becomes one-dimensional, and is unable to solve the problems of modern life. No amount of new information can replace tribal wisdom, which provides the foundation for any good and decent life.

None of what has been said here denies the concept of progress, but it means that genuine progress is the outcome of deep mental dialogue in which new ideas are accepted or rejected by reference to that great complex of old ideas that have been perfected and passed down from one generation to the next generation over many millennia. In other words, genuine progress builds on basic truth. This is not idealization of primitive culture, but conscious recognition of its solid, intelligent achievements. Because civilization repudiates primitive, basic truth, we have no frame of reference for a good and decent life. What we call progress in the modern world is the aimless and reckless rampage of lost individuals. When one is lost, it is necessary to go back to the place where one had one's bearings, and start anew from there.

The False Self

We have internalized our masters, which is a well-known psychological response to trauma. When faced with overwhelming terror, the human mind splits, with part of itself modeling itself after the oppressor. This is an act of appeasement: "Look," the mind says in effect, "I am like you, so do not harm me." As a result of the civilizing process, together with this psychological defense mechanism known as "identification with the aggressor," we now hear the alien voices of the various representatives of civilization in our heads. Because of these alien ego-identifications we no longer hear our own tribal/primal voice. In order for deep thinking to commence again in the human mind, it is necessary to break down these internal authorities, overcome the resistances, that prevent tribal ideas from coming to consciousness. The modern problem is not simply that we do not listen to primal ideas, but rather that primal ideas are unable to come to consciousness at all, because of the internal counterforces, or ego-alien identifications, that contradict and overpower them.

These ego-alien identifications, built up over the course of a lifetime, cohere and form a distinct, circumscribed personality, or false self, that represents and enforces the rules and regulations of civilization. This false self is observable in the frozen facial expressions, stereotypic gestures, and unexamined behavioral patterns of the general public. This false self determines much of our everyday lives, so that we are seldom the origin of our actions. We lapse into the false self at the first sign of danger, under stress, or simply because it is the path of least resistance. In this unthinking mode of social role playing, we internally reproduce our own oppression.

Trauma is a necessary part of civilizing someone, because a natural, maturing individual will not otherwise accept the ideals of civilization. These ideals—hierarchy, property, the State, for example—are so contrary to our tribal nature that they must be forcibly thrust into the human mind. This causes the mind to rupture, to divide its territory—that is, to surrender a part of itself to the invading enemy. For this reason, the false self is never really integrated into the human mind, but instead occupies the mind, as a foreign body, standing apart from and above normal/healthy mental life.

The Original Self

Beneath the false self, there still lives the original identity of the person. This original self is older than, and other than, the foreign personality that has been imposed upon it. This original self or primitive ego is the person one was in childhood—before the mind was ruptured by the trauma of civilization—and the person one still is at the core of one's personal identity. This original self is closely connected to the oldest layer of the psyche. It is an individually differentiated portion of the primitive mind—the first, personal organization of the primitive mind. As such, it has direct access to primal wisdom, an internally directed perceptivity, that permits the entry into consciousness, as well as the passage into activity, of tribal ideas.

In civilization, this original self is caught in the middle of a war between the status quo false self and the uprising of the tribal self.

Even when an individual succeeds in holding its own personal identity, refusing to slip into the false self, the primal voice may remain unheard, smothered by the false self. As a distinct, circumscribed personality, the false self achieves a counter-will of its own that is always operational in the sick mind, at least in terms of repressing tribal ideas. Tribal ideas threaten civilization, so they cannot pass through the censorship of the false self, which functions specifically to ward off all serious challenges to civilization.

Furthermore, the false self tends to become more autonomous and extensive, owing to improved forms of social control and manipulation of consciousness in the modern world—advertising, for example. Once the mind is broken, the false self, backed by the relentless power of civilization, takes over our lives. The original or primary self has become autistic, or severely withdrawn from active engagement with social reality. However, there remains some awareness, at least sometimes, of our primary identity beneath the false self, and so there is hope for lifting the repressions—set down most effectively in childhood—on primitive mental processes.

The True Self

Through identification, which is a normal psychological basis of personality development, the original identity of the person is stretched into something more and greater, on its path to the true self. This process of identification becomes pathological only when the continuity of the personality is not preserved, i.e., when there is a splitting in two of the mind, giving rise to a new psychic structure—a false self. This rupture of the mind is unfortunately precisely what occurs through compulsory training and education of contemporary civilized human beings. In contrast, the development of the natural and sane human mind does not entail any sharp delimitations between its various stages and functions. The true self is a continuation and culmination of the original self; it develops out of the original self, just as the original self develops out of the tribal self. In sum, psychical reality consists of the tribal self, together with the original self, and the dynamic interplay between them is responsible for the development of the true self. The true self is something that unfolds/progresses. Traumatic civilization has not eliminated the original self, but it has stopped the flow of tribal ideas into consciousness, which arrests or distorts the higher/maturer development of the personality. In order to resume deep thinking in the human mind, so that we can become our true self, it is necessary to push back and ultimately dissolve civilization, internally and externally.

The Message

The sadness of the modern age is that the public needs to be reminded of "unchangeable human needs." This truth is not something to be imposed upon them, but something to be evoked within them. The fact is that people do not feel satisfied in the roles assigned to them by civilization. There is a widespread feeling that one's true identity or potential is not being fulfilled, but unfortunately there is no awareness of the tribal self—except among a small minority of individuals. Once the message of a tribal self is re-introduced into public consciousness, it may lie dormant in the minds of individuals for a long time, but it is never entirely forgotten again, as it was in childhood. This message is the catalyst for an intellectual awakening among the population, accompanied by the feeling that something old and familiar has been uncovered.

The power of this message to move an individual is due to the psychological fact that, although repression shuts down deep thinking, tribal ideas continue to push for entrance into consciousness. The mind seeks always to integrate all its ideas into one comprehensible whole. Whatever is part of the repressed unconscious is trying to penetrate into consciousness. When an individual gets the idea of a tribal self from an external source, via the message, it takes hold deep down. The message appeals to the conscious mind, tugs at it, rings true because it touches and stirs the repressed truth that is striving to break into awareness. For this reason, the triumph of civilization over the tribal self is never secure, so there must be a constant bombardment of lies and distortions from the representatives of civilization.

Whether or not an individual awakens to the message of a tribal self depends on the mental condition of one's personal life. Crisis can open a person to ideas that would otherwise be shunned or rejected. A desperate or confused person seeking help may accept the notion of a tribal self intellectually, because it makes sense of one's life-history; but this is not yet to grasp tribal ideas from within, the result of deep mental dialogue. Intellectual awareness of one's tribal self, via the message, is the first prerequisite of a sane person, but it is not enough, not the same thing as hearing one's own primal voice. Each individual must validate or prove, for itself, the message or theory of a tribal self.

The Journey Outward

What is required in order to hear one's own primal voice and resume deep thinking is a journey outward by the original self, which is a matter of fighting to live according to fundamental ideas that one recognizes as good and right, a tribal ideal. This is the path of a warrior because civilization without (laws, institutions, and technologies) and within (the false self) is organized against it. It takes inner resolve and courage to go against civilization, outwardly or inwardly. The path of a tribal warrior in the modern world has nothing to do with picking up a spear or wearing a loin cloth; rather, it involves committing to, and standing up for the great ideas of prehistory: face-to-face democracy, rivers and streams as drinking places, respect for wildlife, etc. These ideas do not require more data, arguments, congresses or conferences because they are the irrefutable elements of original knowledge. The warrior stands on the unshakable foundation of basic truth, and the vicious lies of civilization—that the Earth is property, or that the common good results from each person maximizing its own self-interest, for example—are dismissed as the meaningless babble of the ignorant....

The psychodynamics of the cure consists in overcoming our fear of deep thinking by strengthening the original ego, through everyday practice, to the point that it no longer turns away from its own subversive ideas. We are afraid to think deeply or critically, due to the horror of the past. It is safer not to think about tribal ideas, and spare oneself the painful memories of punishment (and the attacks of anxiety) that are associated with the recall of these ideas. The childhood fear that one will be murdered or grossly harmed for primitive thinking lives on....

The child settles for an unsatisfactory adaptation to social reality—namely, repression of its tribal self because it is unable to deal with the threatening menace of civilization in any other way. However, adults are no longer in the helpless position of childhood; it is no longer necessary to give up the struggle with civilization. There is plenty of opportunity in our everyday lives to question and refuse humiliating and debilitating authority. What matters, for now, is not that we win this or that battle with external reality, but that we stay true to tribal ideas in the face of civilization, and integrate them into the conduct or character of our lives. This is what slowly brings about a radical change in the personality.

The path of a warrior is that of upholding a tribal ideal—with the therapeutic aim of bringing a strengthened ego into direct and open communication with the tribal mind, which dissolves the false self (and its punishment mechanism).

The personal journey of loosening the grip of the false self by strengthening the primary self is certainly not by itself going to bring about the overthrow of civilization; but it is the subjective pre-condition for building an environmental movement that will achieve this end. The public is firmly in the grip of the false self, which means that a revolution is definitely not on the agenda. The influencing machine of civilization is in our heads, and we must defeat it there first; for it is not to be expected that people brutally held down (for 10,000 years) by counter-revolutionary forces will revolt—until the psychological groundwork of liberation has been adequately prepared. After enough of us have done the hard inner work of self-restoration—that is, re-claimed enough of our sanity, then we will come together in tribal units and smash civilization.

Crossroads

Human unhappiness within civilization is widespread and growing. People feel increasingly empty, anxious, depressed, and angry. Everyone is seeking an answer to serious mental problems. The Chinese ideogram for crisis combines the sign of "danger" with that of "opportunity." This is exactly where we are in history, at the crossroads between two radically different futures. On the one hand, there is the danger of insanity, and on the other hand, the opportunity for a return of tribalism.

The Path to Insanity

Basic, vital needs or tribal needs are not being satisfied in civilization, and this produces frustration, which in turn activates aggression. As civilization "progresses" toward global interlock among new technological forms, we are removed farther and farther from the simple and basic joys of life—sitting in a shaft of sunlight, conversing around a fire, food-sharing, walking, the blue sky overhead—and so frustration and aggression also progress. More and more of a person's concern and compassion for others and the natural world are withdrawing, and re-focusing on the self, in order to neutralize the growing frustration and anger within.

This is the psychological explanation for the culture of selfishness that is underway everywhere in the modern world—the first step on the road to insanity.

The second step, which is now on the horizon, is megalomania, a severe pathological state of consciousness in which the steady withdrawal of love from others and nature gives rise to the deluded mental condition of exalted self-importance. The megalomaniac feels more and more powerful, euphoric, and in control of things (due to an abnormally inflated self-love), while in reality s/he is becoming more and more isolated, impotent, and out of control (due to an excessive loss of love for others and nature). This pathological condition of megalomania is fueled by inner hatred, which is desperately seeking pacification by consuming more and more of a person's available love, but it fails entirely to deal with the root-cause of our mental illness—namely, unmet primitive needs.

If people cannot access the tribal ideas that inform them of basic needs, then they cannot find the proper target—namely, civilization—for their frustration and aggression. As a result, rage is trapped in our minds and bodies and is destined to reach heights that are psychologically unbearable for any individual, which leads to madness. Insanity, the third and final stage of civilization, occurs when the original self becomes a tortured prisoner within the walls of its own fears, frustrations, and hostility. It is now extremely difficult, if not impossible, to reach and influence the original self with any rational message.

The Path to Sanity

Civilization has enslaved us in the chains of dependency, isolation, and artificiality. All that we have suffered in civilization still exists in our minds—for nothing vanishes from the human mind—and it is accumulating into rage. The mechanism that civilization employs in order to prevent this rage from discharging itself against its source—namely, civilization—is the establishment of a dictatorship in our heads, a false self, that re-directs this rage toward the citizen in the form of self-blame. According to the false self, each individual is to blame for its own misery. The false self gets the energy it needs to punish and enslave us from our own blind frustration and aggression. Without conscious commitment to a tribal ideal, a firm hold on basic truth, by which to evaluate, condemn, and go against civilization, a person is at the mercy of its own mounting disappointment and aggression that the false self is designed to turn inward.

The tribal ideal is a staff for strengthening the original self and fighting off the false self. By upholding a tribal ideal, sometimes through the spoken word, sometimes through action, and sometimes through silence, we build self-respect on a solid foundation; for at the core of our being we are still fiercely independent, intelligent creatures, with strong affective ties to small human groups and large natural/wild places.

The importance of this self-respect as a prime motivator of human nature can hardly be overstated; it may enable an individual to defy civilization, even in the face of the hatred of the whole modern world. When an individual acquires fundamental self-respect, then s/he will be made a fool no longer, and all the blows of civilization are nothing but the battle scars of a proud warrior. Civilization is powerless against it, because a person who has reclaimed fundamental self-respect cares nothing about the laws and standards of civilization.

This self-respect leads to genuine self-love, the second and decisive step on the path to sanity, for self-love (and happiness in large measure) consists in becoming one's own ideal again, as in childhood. This self-love eventually overflows and becomes love for others and for external nature. Concern for life on Earth is the result of a surplus of love, or as Nietzsche put it, "abundance in oneself ... the over-great fullness of life the feeling of plenitude and increased energy." In other words, caring for life on Earth flows from an unbroken and expanding primary and healthy narcissism. Without this caring that flows from self-love, science and ethics will preach in vain for the preservation of biological diversity, ecological integrity, and real wilderness.

The megalomaniac or pathological narcissist has no love for others or for nature, because s/he needs all available psychic energy, and even more, in order to stave off a false self that is becoming more demanding and brutal. In megalomania, the backward flow of love, away from others and nature and toward the self, is a defensive reaction to the underlying reality of intensifying personal hurt and hatred: the original self is becoming weaker, and the false self stronger, because the gross injury to the psyche of the human being, by the trauma of civilization, festers.

Identification with the tribal ideal is the antidote to the massive narcissistic wounds inflicted on us by civilization, and it threatens civilization with disintegration because it goes to the source of our injury and begins to heal it. The third and final stage of the making of a revolutionary or eco-radical occurs when an individual, after arduous intellectual praxis, either remembers or intuits basic truth. The individual now hears its own primal voice, which rises up from within as an unstoppable conscious drive to assist fellow creatures and to make a positive contribution to the planet.

In sum, the path to sanity begins with the awareness of a tribal self. A personal commitment to this ideal builds self-respect, which in turn builds self-love. This self-love eventually overflows to take in others and nature. At last, individuals are psychologically secure and strong enough to enter into cooperative associations with one another in favor of a mass movement aimed at re-establishing small human villages, embedded in a healthy/wild landscape.

Conclusion

When the final crisis of civilization comes in the 21st century, the present system will do whatever is necessary to perpetuate itself. People who are in the stranglehold of the false self will support whatever the system does, including the exploitation and destruction of National Parks, Designated Wilderness Areas, and The Wildlands Project (whatever it achieves). Unless people possess an assured, sane psychological core, everything else is in jeopardy. The personal pursuit of sanity is, or will shortly become, the overriding issue for the vast majority of people, and toward this end, each person, isolated and fearful in modern society, must be armed with an ideal or vision of what s/he would like to be and should be. There is nothing better to offer people than a tribal ideal. It is, then, up to each person to do the difficult psychological work of personal liberation that must precede real change. This personal task of healing the mind can be and must be accomplished, to a decisive point, in modern-day circumstances.

"Revolt of the Bats" (1995)

Alon K. Raab

> *North America, Turtle Island, taken*
> *by invaders who wage war*
> *on the world,*
>
> *May ants, may abalone, otters,*
> *wolves, and elk rise!*
> *and pull away their giving*
> *from the robot nations.*
>
> —Gary Snyder
> **Mother Earth: Her Whales**

The animals are fighting back. By tooth and claw, by wing and paw, they are waging a war against civilized tyranny and destruction.

Sympathetic humans are burning down farm and fur ranch equipment, demolishing butcher shops, and trying to stop rodeos, circuses, and other forms of "entertainment." But the animals are also acting as their own defenders, fighting for their own liberation.

These actions of revolt are done by individual animals, as well as by whole communities, and take many forms. Escape from captivity is a commonly employed tactic.

Here I would like to remember and salute the orangutan who escaped from his prison cell at the Kansas City Zoo in June 1990 by unscrewing four large bolts; the West African Cape clawless otter who, in December 1991, pushed her way through the wired cage at the Portland zoological incarceration facilities; an alligator who climbed a high ramp at a Seattle science exhibition in October 1991 and vanished for several hours; the elephant at the Louisville Zoo who escaped in June 1994; the sea otter "Cody" who in September 1993, armed with a fiberglass bolt pried from the floor of the Oregon Coast Aquarium took aim at a window and shattered one of the glass layers; the chimpanzees "Ai" and "Akira" at the Kyoto University Primates Research institute, who used keys taken from a guard to open their cages, cross the hall to free their orangutan friend "Doodoo," and bolt to freedom.

In April 1990, a cow destined for a Turkish slaughterhouse leapt from the truck onto the roof of a car carrying a provincial governor, crushing it and injuring the official. The fate of the cow was not reported, but one hopes she was able to make her way into the hills. A decade earlier, near the town of Salem, Oregon, "Rufus" the bull knocked down the door of a truck carrying him to be butchered, and roamed freely for a few days until captured by bounty hunters, and returned to his "owner." And in Cairo, Egypt, in June 1993, during the Muslim Eid-Al-Adha ("feast of sacrifice"), a bull escaped upon catching a glimpse of the butcher's knife. The animal chased its would-be slaughterer up to his third floor apartment, wrecking furniture and forcing him to hide in the bedroom.

Some of the animals were recaptured and returned to their prisons, but the otter, who was last seen crossing the roadway between the Portland Zoo and the Oregon Museum of Science and Industry, making her way into the nearby forests, is a true inspiration, and hopefully a harbinger of many more daring dashes.

Sometimes free animals are in a strategic position to resist greed and profit. In 1991, a bald eagle blocked plans for a three million dollar road expansion project in Central Oregon by nesting near Highway 20. An eagle standing in the way of motorized mania is a beautiful sight to behold.

There are animals who carry the battle a step further, like the wren, nesting in a Washington, D.C. traffic light, who swooped down to attack business people. Other birds commit suicide by entering military plane engines and decommissioning them. In a show of solidarity for a fellow animal, the mule "Ruthie" kicked her rider, Idaho Governor Cecil Andrus, during a hunting trip, as he was loading a murdered elk onto her. Andrus suffered a broken nose and deep lacerations.

The Belgian spaniel who discharged a shotgun, killing hunter Jean Guillaume, the elephant who gored hunter Alan Lowe in Zimbabwe, and the cow who killed Quebec farmer Origene Ste-beanne when he tried to steal her newborn calf, are also worthy of our respect. I prefer persuasion and education to the taking of life, but there is poetic justice in these accounts.

When animals band together they are able to unleash a mighty power. Several years ago, in the depths of the suburban wastelands of Springfield, Massachusetts, ring-billed gulls bombarded a new golf course and its patrons with golf balls. The shocked golfers were forced to withdraw from their favorite water-and-land-wasting activity for several weeks, and consider the fact that for many years these lands were nesting grounds for the birds.

In the summer of 1989, downtown Fort Worth, Texas, came to a halt when thousands of Mexican free-tail bats descended on the city. In the early years of this century, bats wreaked much havoc on many Texas towns. In Austin, bats invaded the courthouse and Capitol building, flying through court sessions, stopping trials and nesting in the dark and cool buildings.

The bats that appeared in Fort Worth chewed into telephone lines and interrupted business as usual. The bats were a reminder to the local population, encased in glass and steel tombstones known as "offices," that this world is much more complex and wondrous than anything taught in management courses. After a day, the bats vanished as they had come, into the unknown.

In the ancient myths of humanity, a special place of respect is given to animals. Affecting people in mysterious ways, and embodying particular qualities, they acted as messengers, as bearers of souls and gifts, and as symbols of all that was wonderful and magical. Birds, fish and mammals (and their many mutations with humans) were presented in myriad ways. A common theme was their ability to fend off hostile human attacks, through trickery, playfulness and wisdom. Coyote and Raven of the Northwest Coast of Turtle Island, the Keen Keeng of Australian dream time, and the sacred bee of Rhodes, are but some of the many magical beings who protected themselves and the lives of other animals and plants.

Once writing developed, accounts of animals opposing human arrogance and avarice abounded in the literature of natural history. We need only look at the inspiring reports provided by the Roman, Pliny the Elder. He marvels at elephants who trampled hunters, refusing to fight their kin in circuses and attempting to break loose from their shackles. Pliny also wrote of dolphins who rushed to rescue other dolphins from captivity, and of wild horses, loons, oxen, dogfish, rabbits and giant centipedes who resisted humans and often won. His accounts also include many instances of alliances between animals and aware humans, each assisting the other, and gaining mutual love and respect.

The medieval work, *On the Criminal Persecution of Animals*, provides in great detail the legacies of pigs, cows, sparrows, ravens, sheep, mules, horses and even worms, who brought destruction upon the human world. Animals disturbing church services, interrupting religious processions at their most solemn moment, and spoiling food supplies were common occurrences. As ancient traditions celebrating the sanctity of nature were rooted out and replaced by an anti-life world view, these animals were accused of being in league with demonic forces. The Christian courts held them responsible for their actions. The "criminals" were tried in regular courts of law, convicted and severely punished. In their pious zeal, the accusers missed the fact that the two-legged and four-legged beings were engaged in guerrilla warfare. They were revolting against humans who were attacking the rivers, valleys and forests. They were opposing the invaders who were engaged in that process of control, euphemistically called "domestication," which, in reality, is enslavement and ecocide.

We are now living in the age of rationality and science, where well-meaning people feel no shame blurting out clichés like "finding the balance between the environment and economic interests," or "managing wildlife," as if wilderness was a commodity to profit from, control and manipulate.

The destruction of the wild (out there, and in our own souls) proceeds at an ever-maddening pace. Let us hope that acts of self-defense and resistance by animals, fish, birds and their human brothers and sisters increases. Let these actions multiply and intensify until human tyranny is thrown off and replaced by a community of free living beings, assisting each other in this magical journey, and reforging the ancient bonds of beauty and camaraderie.

Rebels Against the Future: Lessons from the Luddites (1995)

Kirkpatrick Sale

Industrial civilization is today the water we swim in, and we seem almost as incapable of imagining what an alternative might look like, or even realizing that an alternative could exist, as fish in the ocean.

The political task of resistance today, then—beyond the "quiet acts" of personal withdrawal Mumford urges—is to try to make the culture of industrialism and its assumptions less invisible and to put the issue of its technology on the political agenda, in industrial societies as well as their imitators. In the words of Neil Postman, a professor of communications at New York University and author of *Technopoly*, "it is necessary for a great debate" to take place in industrial society between "technology and everybody else" around all the issues of the "uncontrolled growth of technology" in recent decades. This means laying out as clearly and fully as possible the costs and consequences of our technologies, in the near term and long, so that even those overwhelmed by the ease/comfort/speed/power of high-tech gadgetry (what Mumford called technical "bribery") are forced to understand at what price it all comes and who is paying for it. What purpose does this machine serve? What problem has become so great that it needs this solution? Is this invention nothing but, as Thoreau put it, an improved means to an unimproved end? It also means forcing some awareness of who the principal beneficiaries of the new technology are—they tend to be the large, bureaucratic, complex, and secretive organizations of the industrial world—and trying to make public all the undemocratic ways they make the technological choices that so affect all the rest of us. Who are the winners, who the losers? Will this concentrate or disperse power, encourage or discourage self-worth? Can society at large afford it? Can the biosphere?

Ultimately this "great debate" of course has to open out into wider questions about industrial society itself, its values and purposes, its sustainability. It is no surprise that the Luddites were unable to accomplish this in the face of an immensely self-satisfied laissez-faire plutocracy whose access to means of forcing debates and framing issues was considerably greater than theirs. Today, though, that task ought not to be so difficult—in spite of the continued opposition of a plutocracy grown only more powerful and complacent—particularly because after two centuries it is now possible to see the nature of industrial civilization and its imperiling direction so much more clearly.

Certain home-truths are beginning to be understood, at least in most industrial societies, by increasing numbers of people: some of the fish at least not only seem to be seeing the water but realizing it is polluted. Industrialism, built upon machines designed to exploit and produce for human betterment alone, is on a collision course with the biosphere. Industrial societies, which have shown themselves capable of creating material abundance for a few and material improvement for many, are nonetheless shot through with inequality, injustice, instability, and incivility, deficiencies that seem to increase rather than decrease with technical advancement. Industrialism does not stand superior, on any level other than physical comfort and power and a problematic longevity of life, to many other societies in the long range of the human experiment, particularly those, morally based and earth-regarding, that did serve the kind of "apprenticeship to nature" that Herbert Read saw as the proper precondition to technology.

Say what you will about such tribal societies, the record shows that they were (and in some places still are) units of great cohesion and sodality, of harmony and regularity, devoid for the most part of crime or addiction or anomie or poverty or suicide, with comparatively few needs and those satisfied with a minimum of drudgery, putting in on average maybe four hours a day per person on tasks of hunting and gathering and cultivating, the rest of the time devoted to song and dance and ritual and sex and eating and stories and games.... No, they did not have the power of 500 servants at the flick of a switch or turn of a key, but then they did not have atomic bombs and death camps, toxic wastes, traffic jams, strip mining, organized crime, psychosurgery, advertising, unemployment or genocide.

To propose, in the midst of the "great debate," that such societies are exemplary, instructive if not imitable, is not to make a romanticized "search for the primitive." It is rather to acknowledge that the tribal mode of existence, precisely because it is nature-based, is consonant with the true, underlying needs of the human creature, and that we denigrate that mode and deny those needs to our loss and disfigurement. It is to suggest that certain valuable things have been left behind as we have sped headlong down the tracks of industrial progress and that it behooves us, in a public and spirited way, to wonder about what we have gained from it all and reflect upon what we have lost. And it is, finally, to assert that some sort of ecological society, rooted in that ancient animistic, autochthonous tradition, must be put forth as the necessary, achievable goal for human survival and harmony on earth.

Philosophically, resistance to industrialism must be embedded in an analysis—an ideology, perhaps—that is morally informed, carefully articulated, and widely shared.

One of the failures of Luddism (if at first perhaps one of its strengths) was its formlessness, its unintentionality, its indistinctness about goals, desires, possibilities. Movements acting out of rage and outrage are often that way, of course, and for a while there is power and momentum in those alone. For durability, however, they are not enough, they do not sustain a commitment that lasts through the adversities of repression and trials, they do not forge a solidarity that prevents the infiltration of spies and stooges, they do not engender strategies and tactics that adapt to shifting conditions and adversaries, and they do not develop analyses that make clear the nature of the enemy and the alternatives to put in its place.

Now it would be difficult to think that neo-Luddite resistance, whatever form it takes, would be able to overcome all those difficulties, particularly on a national or international scale: commitment and solidarity are mostly products of face-to-face, day-to-day interactions, unities of purpose that come from unities of place. But if it is to be anything more than sporadic and martyristic, resistance could learn from the Luddite experience at least how important it is to work out some common analysis that is morally clear about the problematic present and the desirable future, and the common strategies that stem from it.

All the elements of such an analysis, it seems to me, are in existence, scattered and still needing refinement, perhaps, but there: in Mumford and Schumacher and Wendell Berry and Jerry Mander and the Chellis Glendinning manifesto; in the writing of the Earth Firsters and the bioregionalists and deep ecologists; in the lessons and models of the Amish and the Irokwa; in the wisdom of tribal elders and the legacy of tribal experience everywhere; in the work of the long line of dissenters-from-progress and naysayers-to-technology. I think we might even be able to identify some essentials of that analysis, such as:

Industrialism, the ethos encapsulating the values and technologies of Western civilization, is seriously endangering stable social and environmental existence on this planet, to which must be opposed the values and techniques of an organic ethos that seeks to preserve the integrity, stability, and harmony of the biotic community, and the human community within it.

Anthropocentrism, and its expression in both humanism and monotheism, is the ruling principle of that civilization, as to which must be opposed the principle of biocentrism and the spiritual identification of the human with all living species and systems.

Globalism, and its economic and military expression, is the guiding strategy of that civilization, to which must be opposed the strategy of localism, based upon the empowerment of the coherent bioregion and the small community.

Industrial capitalism, as an economy built upon the exploitation and degradation of the earth, is the productive and distributive enterprise of that civilization, to which must be opposed the practices of an ecological and sustainable economy built upon accommodation and commitment to the earth and following principles of conservation, stability, self-sufficiency, and cooperation.

A movement of resistance starting with just those principles as the sinews of analysis would at least have a firm and uncompromising ground on which to stand and a clear and inspirational vision of where to go. If nothing else, it would be able to live up to the task that George Grant, the Canadian philosopher, has set this way: "The darkness which envelops the Western world because of its long dedication to the overcoming of chance"—by which he means the triumph of the scientific mind and its industrial constructs—"is just a fact. The job of thought in our time is to bring into the light that darkness as darkness." And at its best, it might bring into the light the dawn that is the alternative....

If the edifice of industrial civilization does not eventually crumble as a result of a determined resistance within its very walls, it seems certain to crumble of its own accumulated excesses and instabilities within not more than a few decades, perhaps sooner, after which there may be space for alternative societies to arise.

The two chief strains pulling this edifice apart, environmental overload and social dislocation, are both the necessary and inescapable results of an industrial civilization. In some sense, to be sure, they are the results of *any* civilization: the record of the last five thousand years of history clearly suggests that every single preceding civilization has perished, no matter where or how long it has been able to flourish, as a result of its sustained assault on its environment, usually ending in soil loss, flooding, and starvation, and a successive distension of all social strata, usually ending in rebellion, warfare, and dissolution. Civilizations, and the empires that give them shape, may achieve much of use and merit—or so the subsequent civilization's historians would have us believe—but they seem unable to appreciate scale or limits, and in their growth and turgidity cannot maintain balance and continuity within or without. Industrial civilization is different only in that it is now much larger and more powerful than any known before, by geometric differences in all dimensions, and its collapse will be far more extensive and thoroughgoing, far more calamitous.

It is possible that such a collapse will be attended by environmental and social dislocations so severe that they will threaten the continuation of life, at least human life, on the surface of the planet, and the question then would be whether sufficient numbers survive and the planet is sufficiently hospitable for scattered human communities to emerge from among the ashes. But it is also possible that it will come about more by decay and distension, the gradual erosion of nation-state arrangements made obsolete and unworkable, the disintegration of corporate behemoths unable to comprehend and respond, and thus with the slow resurrection and re-empowerment of small bioregions and coherent communities having control over their own political and economic destinies. In either case, it will be necessary for the survivors to have some body of lore, and some vision of human regeneration, that instructs them in how thereafter to live in harmony with nature and how and why to fashion their technologies with the restraints and obligations of nature intertwined, seeking not to conquer and dominate and control the species and systems of the natural world—for the failure of industrialism will have taught the folly of that—but rather to understand and obey and love and incorporate nature into their souls as well as their tools.

It is now the task of the neo-Luddites, armed with the past, to prepare, to preserve, and to provide that body of lore, that inspiration, for such future generations as may be.

"ACTIONS SPEAK LOUDER THAN WORDS" (1998)

DERRICK JENSEN

Every morning when I wake up I ask myself whether I should write or blow up a dam. I tell myself I should keep writing, though I'm not sure that's right. I've written books and done activism, but it is neither a lack of words nor a lack of activism that is killing salmon here in the Northwest. It's the dams.

Anyone who knows anything about salmon knows the dams must go. Anyone who knows anything about politics knows the dams will stay. Scientists study, politicians and business people lie and delay, bureaucrats hold sham public meetings, activists write letters and press releases, and still the salmon die.

Sadly enough, I'm not alone in my inability or unwillingness to take action. Members of the German resistance to Hitler from 1933 to 1945, for example, exhibited a striking blindness all too familiar: Despite knowing that Hitler had to be removed for a "decent" government to be installed, they spent more time creating paper versions of this theoretical government than attempting to remove him from power. It wasn't a lack of courage that caused this blindness but rather a misguided sense of morals. Karl Goerdeler, for instance, though tireless in attempting to create this new government, staunchly opposed assassinating Hitler, believing that if only the two could sit face to face Hitler might relent.

We, too, suffer from this blindness and must learn to differentiate between real and false hopes. We must eliminate false hopes, which blind us to real possibilities. Does anyone really believe our protests will cause Weyerhaeuser or other timber transnationals to stop destroying forests? Does anyone really believe the same corporate administrators who say they "wish salmon would go extinct so we could just get on with living" (Randy Hardy of Bonneville Power Association) will act other than to fulfill their desires? Does anyone really believe a pattern of exploitation as old as our civilization can be halted legislatively, judicially or through means other than an absolute rejection of the mindset that engineers the exploitation, followed by actions based on that rejection? Does anybody really think those who are destroying the world will stop because we ask nicely or because we lock arms peacefully in front of their offices?

There can be few who still believe the purpose of government is to protect citizens from the activities of those who would destroy. The opposite is true: Political economist Adam Smith was correct in noting that the primary purpose of government is to protect those who run the economy from the outrage of injured citizens. To expect institutions created by our culture to do other than poison waters, denude hillsides, eliminate alternative ways of living and commit genocide is unforgivably naive.

Many German conspirators hesitated to remove Hitler from office because they'd sworn loyalty to him and his government. Their scruples caused more hesitation than their fear. How many of us have yet to root out misguided remnants of a belief in the legitimacy of this government to which, as children, we pledged allegiance? How many of us fail to cross the line into violent resistance because we still believe that, somehow, the system can be reformed? And if we don't believe that, what are we waiting for? As Shakespeare so accurately put it, "Conscience doth make cowards of us all."

It could be argued that by comparing our government to Hitler's I'm overstating my case. I'm not sure salmon would agree, nor lynx, nor the people of Peru, Irian Jaya, Indonesia, or any other place where people pay with their lives for the activities of our culture.

If we're to survive, we must recognize that we kill by inaction as surely as by action. We must recognize that, as Hermann Hesse wrote, "We kill when we close our eyes to poverty, affliction or infamy. We kill when, because it is easier, we countenance, or pretend to approve of atrophied social, political, educational, and religious institutions, instead of resolutely combating them."

The central—and in many ways only—question of our time is this: What are sane, appropriate and effective responses to outrageously destructive behavior? So often, those working to slow the destruction can plainly describe the problems. Who couldn't? The problems are neither subtle nor cognitively challenging. Yet when faced with the emotionally daunting task of fashioning a response to these clearly insoluble problems, we generally suffer a failure of nerve and imagination. Gandhi wrote a letter to Hitler asking him to stop committing atrocities and was mystified that it didn't work. I continue writing letters to the editor of the local corporate newspaper pointing out mistruths and am continually surprised at the next absurdity.

I'm not suggesting a well-targeted program of assassinations would solve all of our problems. If it were that simple, I wouldn't be writing this essay. To assassinate Slade Gorton and Larry Craig, for example, two senators from the Northwest whose work may be charitably described as unremittingly ecocidal, would probably slow the destruction not much more than to write them a letter. Neither unique nor alone, Gorton and Craig are merely tools for enacting ecocide, as surely as are dams, corporations, chainsaws, napalm and nuclear weapons. If someone were to kill them, others would take their places. The ecocidal programs originating specifically from the damaged psyches of Gorton and Craig would die with them, but the shared nature of the impulses within our culture would continue full-force, making the replacement as easy as buying a new hoe.

Hitler, too, was elected as legally and "democratically" as Craig and Gorton. Hitler, too, manifested his culture's death urge brilliantly enough to capture the hearts of those who voted him into power and to hold the loyalty of the millions who actively carried out his plans. Hitler, like Craig and Gorton, like George Weyerhaeuser and other CEOs, didn't act alone. Why, then, do I discern a difference between them?

The current system has already begun to collapse under the weight of its ecological excesses, and here's where we can help. Having transferred our loyalty away from our culture's illegitimate economic and governmental entities and to the land, our goal must be to protect, through whatever means possible, the human and nonhuman residents of our homelands. Our goal, like that of a demolition crew on a downtown building, must be to help our culture collapse in place, so that in its fall it takes out as little life as possible.

Discussion presupposes distance, and the fact that we're talking about whether violence is appropriate tells me we don't yet care enough. There's a kind of action that doesn't emerge from discussion, from theory, but instead from our bodies and from the land. This action is the honeybee stinging to defend her hive; it's the mother grizzly charging a train to defend her cubs; it's Zapatista spokesperson Cecelia Rodriguez saying, "I have a question of those men who raped me. Why did you not kill me? It was a mistake to spare my life. I will not shut up ... this has not traumatized me to the point of paralysis." It's Ogoni activist Ken Saro-Wiwa, murdered by the Nigerian government at the urging of Shell, whose last words were, "Lord, take my soul, but the struggle continues!" It's those who participated in the Warsaw Ghetto uprising. It's Crazy Horse, Sitting Bull and Geronimo. It's salmon battering themselves against concrete, using the only thing they have, their flesh, to try to break down that which keeps them from their homes.

I don't believe the question of whether to use violence is the right one. Instead, the question should be: Do you sufficiently feel the loss? So long as we discuss this in the abstract, we still have too much to lose. If we begin to feel in our bodies the immensity and emptiness of what we lose daily—intact natural communities, hours sold for wages, childhoods lost to violence, women's capacity to walk unafraid—we'll know precisely what to do.

"We Have To Dismantle All This" (1995)

Anti-Authoritarians Anonymous

The unprecedented reality of the present is one of enormous sorrow and cynicism, "a great tear in the human heart," as Richard Rodriguez put it. A time of ever-mounting everyday horrors, of which any newspaper is full, accompanies a spreading environmental apocalypse. Alienation and the more literal contaminants compete for the leading role in the deadly dialectic of life in divided, technology-ridden society. Cancer, unknown before civilization, now seems epidemic in a society increasingly barren and literally malignant.

Soon, apparently, everyone will be using drugs; prescription and illegal becoming a relatively unimportant distinction. Attention Deficit Disorder is one example of an oppressive effort to medicalize the rampant restlessness and anxiety caused by a life-world ever more shriveled and unfulfilling. The ruling order will evidently go to any lengths to deny social reality; its techno-psychiatry views human suffering as chiefly biological in nature and genetic in origin.

New strains of disease, impervious to industrial medicine, begin to spread globally while fundamentalism (Christian, Judaic, Islamic) is also on the rise, a sign of deeply-felt misery and frustration. And here at home New Age spirituality (Adorno's "philosophy for dunces") and the countless varieties of "healing" therapies wear thin in their delusional pointlessness. To assert that we can be whole/enlightened/healed within the present madness amounts to endorsing the madness.

The gap between rich and poor is widening markedly in this land of the homeless and the imprisoned. Anger rises and massive denial, cornerstone of the system's survival, is now at least having a troubled sleep. A false world is beginning to get the amount of support it deserves: distrust of public institutions is almost total. But the social landscape seems frozen and the pain of youth is perhaps the greatest of all. It was recently announced (10/94) that the suicide rate among young men ages 15 to 19 more than doubled between 1985 and 1991. Teen suicide is the response of a growing number who evidently cannot imagine maturity in such a place as this.

The overwhelmingly pervasive culture is a fast-food one, bereft of substance or promise. As Dick Hebdige aptly judged, "the postmodern is the modern without the hopes and dreams that made modernity bearable." Postmodernism advertises itself as pluralistic, tolerant, and non-dogmatic. In practice it is a superficial, fast-forward, deliberately confused, fragmented, media-obsessed, illiterate, fatalistic, uncritical excrescence, indifferent to questions of origins, agency, history or causation. It questions nothing of importance and is the perfect expression of a setup that is stupid and dying and wants to take us with it.

Our postmodern epoch finds its bottom-line expression in consumerism and technology, which combine in the stupefying force of mass media. Attention-getting, easily-digested images and phrases distract one from the fact that this horror-show of domination is precisely held together by such entertaining, easily digestible images and phrases. Even the grossest failures of society can be used to try to narcotize its subjects, as with the case of violence, a source of endless diversion. We are titillated by the representation of what at the same time is threatening, suggesting that boredom is an even worse torment than fear.

Nature, what is left of it, that is, serves as a bitter reminder of how deformed, non-sensual, and fraudulent is contemporary existence. The death of the natural world and the technological penetration of every sphere of life, what is left of it, proceed with an accelerating impetus. *Wired*, *Mondo 2000*, zippies, cyber-everything, virtual reality, Artificial Intelligence, on and on, up to and including Artificial Life, the ultimate postmodern science.

Meanwhile, however, our "post-industrial" computer age has resulted in the fact that we are *more than ever* "appendages to the machine," as the 19th-century phrase had it. Bureau of Justice statistics (7/94), by the way, report that the increasingly computer-surveilled workplace is now the setting for nearly one million violent crimes per year, and that the number of murdered bosses has doubled in the past decade.

This hideous arrangement expects, in its arrogance, that its victims will somehow remain content to vote, recycle, and pretend it will all be fine. To employ a line from Debord, "The spectator is simply supposed to know nothing and deserve nothing."

Civilization, technology, and a divided social order are the components of an indissoluble whole, a death-trip that is fundamentally hostile to qualitative difference. Our answer must be qualitative, not the quantitative, more-of-the-same palliatives that actually reinforce what we must end.

Talking to the Owls
and Butterflies (1976)

John (Fire) Lame Deer and Richard Erdoes

Let's sit down here, all of us, on the open prairie, where we can't see a highway or a fence. Let's have no blankets to sit on, but feel the ground with our bodies, the earth, the yielding shrubs. Let's have the grass for a mattress, experiencing its sharpness and its softness. Let us become like stones, plants, and trees. Let us be animals, think and feel like animals.

Listen to the air. You can hear it, feel it, smell it, taste it. *Woniya waken*—the holy air—which renews all by its breath. *Woniya, woniya waken*—spirit, life, breath, renewal—it means all that. *Woniya*—we sit together, don't touch, but something is there; we feel it between us, as a presence. A good way to start thinking about nature, talk about it. Rather talk about it, talk to the rivers, to the lakes, to the winds as to our relatives.

You have made it hard for us to experience nature in the good way by being part of it. Even here we are conscious that somewhere out in those hills there are missile silos and radar stations. White men always pick the few unspoiled, beautiful, awesome spots for the sites of these abominations. You have raped and violated these lands, always saying, "Gimme, gimme, gimme," and never giving anything back. You have taken 200,000 acres of our Pine Ridge reservation and made them into a bombing range. This land is so beautiful and strange that now some of you want to make it into a national park. The only use you have made of this land since you took it from us was to blow it up. You have not only despoiled the earth, the rocks, the minerals, all of which you call "dead" but which are very much alive; you have even changed the animals, which are part of us, part of the Great Spirit, changed them in a horrible way, so no one can recognize them. There is power in a buffalo—spiritual, magic power—but there is no power in an Angus, in a Hereford.

There is power in an antelope, but not in a goat or in a sheep, which holds still while you butcher it, which will eat your newspaper if you let it. There was great power in a wolf, even in a coyote. You have made him into a freak—a toy poodle, a Pekingese, a lap dog. You can't do much with a cat, which is like an Indian, unchangeable. So you fix it, alter it, declaw it, even cut its vocal cords so you can experiment on it in a laboratory without being disturbed by its cries.

A partridge, a grouse, a quail, a pheasant, you have made them into chickens, creatures that can't fly, that wear a kind of sunglasses so that they won't peck each other's eyes out, "birds" with a "pecking order." There are farms where they breed chickens for breast meat. Those birds are kept in low cages, forced to be hunched over all the time, which makes the breast muscles very big. Soothing sounds, Muzak, are piped into these chicken hutches. One loud noise and the chickens go haywire, killing themselves by flying against the mesh of their cages. Having to spend all their lives stooped over makes an unnatural, crazy, no-good bird. It also makes unnatural, no-good human beings.

That's where you fooled yourselves. You have not only altered, declawed and malformed your winged and four-legged cousins; you have done it to yourselves. You have changed men into chairmen of boards, into office workers, into time-clock punchers. You have changed women into housewives, truly fearful creatures. I was once invited into the home of such a one.

"Watch the ashes, don't smoke, you stain the curtains. Watch the goldfish bowl, don't breathe on the parakeet, don't lean your head against the wallpaper; your hair may be greasy. Don't spill liquor on that table: it has a delicate finish. You should have wiped your boots; the floor was just varnished. Don't, don't, don't..." That is crazy. We weren't made to endure this. You live in prisons which you have built for yourselves, calling them "homes," offices, factories. We have a new joke on the reservation: "What is cultural deprivation?" Answer: "Being an upper-middle-class white kid living in a split-level suburban home with a color TV."

Sometimes I think that even our pitiful tar-paper shacks are better than your luxury homes. Walking a hundred feet to the outhouse on a clear wintry night, through mud or snow. That's one small link with nature. Or in the summer, in the back country, leaving the door of the privy open, taking your time, listening to the humming of the insects, the sun warming your bones through the thin planks of wood; you don't even have that pleasure anymore.

Americans want to have everything sanitized. No smells! Not even the good, natural man and woman smell. Take away the smell from under the armpits, from your skin. Rub it out, and then spray or dab some nonhuman odor on yourself, stuff you can spend a lot of money on, ten dollars an ounce, so you know this has to smell good. "B.O.," bad breath, "Intimate Female Odor Spray"—I see it all on TV. Soon you'll breed people without body openings.

I think white people are so afraid of the world they created that they don't want to see, feel, smell or hear it. The feeling of rain and snow on your face, being numbed by an icy wind and thawing out before a smoking fire, coming out of a hot sweat bath and plunging into a cold stream, these things make you feel alive, but you don't want them anymore. Living in boxes which shut out the heat of the summer and the chill of the winter, living inside a body that no longer has a scent, hearing the noise from the hi-fi instead of listening to the sounds of nature, watching some actor on TV having a make-believe experience when you no longer experience anything for yourself, eating food without taste—that's your way. It's no good.

The food you eat, you treat it like your bodies, take out all the nature part, the taste, the smell, the roughness, then put the artificial color, the artificial flavor in. Raw liver, raw kidney—that's what we old-fashioned full-bloods like to get our teeth into. In the old days we used to eat the guts of the buffalo, making a contest of it, two fellows getting hold of a long piece of intestines from opposite ends, starting chewing toward the middle, seeing who can get there first; that's eating. Those buffalo guts, full of half-fermented, half-digested grass and herbs, you didn't need any pills and vitamins when you swallowed those. Use the bitterness of gall for flavoring, not refined salt or sugar. Wasna—meat, kidney fat and berries all pounded together—a lump of that sweet wasna kept a man going for a whole day. That was food, that had the power. Not the stuff you give us today: powdered milk, dehydrated eggs, pasteurized butter, chickens that are all drumsticks or all breast; there's no bird left there.

You don't want the bird. You don't have the courage to kill honestly—cut off the chicken's head, pluck it and gut it—no, you don't want this anymore. So it all comes in a neat plastic bag, all cut up, ready to eat, with no taste and no guilt. Your mink and seal coats, you don't want to know about the blood and pain that went into making them. Your idea of war—sit in an airplane, way above the clouds, press a button, drop the bombs, and never look below the clouds—that's the odorless, guiltless, sanitized way.

When we killed a buffalo, we knew what we were doing. We apologized to his spirit, tried to make him understand why we did it, honoring with a prayer the bones of those who gave their flesh to keep us alive, praying for their return, praying for the life of our brothers, the buffalo nation, as well as for our own people. You wouldn't understand this and that's why we had the Washita Massacre, the Sand Creek Massacre, the dead women and babies at Wounded Knee. That's why we have Song My and My Lai now.

To us life, all life, is sacred. The state of South Dakota has pest-control officers. They go up in a plane and shoot coyotes from the air. They keep track of their kills, put them all down in their little books. The stockmen and sheep owners pay them. Coyotes eat mostly rodents, field mice and such. Only once in a while will they go after a stray lamb. They are our natural garbage men cleaning up the rotten and stinking things. They make good pets if you give them a chance. But their living could lose a man a few cents, and so the coyotes are killed from the air. They were here before the sheep, but they are in the way; you can't make a profit out of them. More and more animals are dying out. The animals which the Great Spirit put here, they must go. The man-made animals are allowed to stay—at least until they are shipped out to be butchered. That terrible arrogance of the white man, making himself something more than God, more than nature, saying, "I will let this animal live because it makes money"; saying, "This animal must go, it brings no income, the space it occupies can be used in a better way. The only good coyote is a dead coyote." They are treating coyotes almost as badly as they used to treat Indians.

You are spreading death, buying and selling death. With all your deodorants, you smell of it, but you are afraid of its reality; you don't want to face up to it. You have sanitized death, put it under the rug, robbed it of its honor. But we Indians think a lot about death. I do. Today would be a perfect day to die—not too hot, not too cool. A day to leave something of yourself behind, to let it linger. A day for a lucky man to come to the end of his trail. A happy man with many friends. Other days are not so good. They are for selfish, lonesome men, having a hard time leaving this earth. But for whites every day would be considered a bad one, I guess.

Eighty years ago our people danced the Ghost Dance, singing and dancing until they dropped from exhaustion, swooning, fainting, seeing visions. They danced in this way to bring back the dead, to bring back the buffalo. A prophet had told them that through the power of the Ghost Dance the earth would roll up like a carpet, with all the white man's works—the fences and the mining tones with their whorehouses, the factories and the farms with their stinking, unnatural animals, the railroads and the telegraph poles, the whole works. And underneath this rolled-up white man's world we would find again the flowering prairies, unspoiled, with its herds of buffalo and antelope, its clouds of birds, belonging to everyone, enjoyed by all.

I guess it was not time for this to happen, but it is coming back, I feel it warming my bones. Not the old Ghost Dance, not the rolling-up—but a new-old spirit, not only among Indians but among whites and blacks, too, especially among young people. It is like raindrops making a tiny brook, many brooks making a stream, many streams making one big river bursting all dams. Us making this book, talking like this—these are some of the raindrops.

Listen, I saw this in my mind not long ago: In my vision the electric light will stop sometime. It is used too much for TV and going to the moon. The day is coming when nature will stop the electricity. Police without flashlights, beer getting hot in the refrigerators, planes dropping from the sky, even the President can't call up somebody on the phone. A young man will come, or men, who'll know how to shut off all electricity. It will be painful, like giving birth. Rapings in the dark, winos breaking into liquor stores, a lot of destruction. People are being too smart, too clever, the machine stops and they are helpless, because they have forgotten how to make do without the machine. There is a Light Man coming, bringing a new light. It will happen before the century is over. The man who has the power will do good things, too—stop all atomic power, stop wars, just by shutting the white electro-power off. I hope to see this, but then I'm also afraid. What will be will be.

Anarcho-Futurist Manifesto (1919)

Group of Anarcho-Futurists

Ah-ah-ah, ha-ha, ho-ho!

Fly into the streets! All who are still fresh and young and not dehumanized—to the streets! The pot-bellied mortar of laughter stands in a square drunk with joy. Laughter and Love, copulating with Melancholy and Hate, pressed together in the mighty, convulsive passion of bestial lust. Long live the psychology of contrasts! Intoxicated, burning spirits have raised the flaming banner of intellectual revolution. Death to the creatures of routine, the philistines, the sufferers from gout! Smash with a deafening noise the cup of vengeful storms! Tear down the churches and their allies the museums! Blast to smithereens the fragile idols of Civilization! Hey, you decadent architects of the sarcophagi of thought, you watchmen of the universal cemetery of books— stand aside! We have come to remove you! The old must be buried, the dusty archive burned by Vulcan's torch of creative genius. Past the flaky ashes of worldwide devastation, past the charred canvases of bulky paintings, past the burned, fat, pot-bellied volumes of classics we march, we Anarcho-Futurists! Above the vast expanse of devastation covering our land the banner of anarchy will be proudly unfurled! Writing has no value! There is no market for literature! There are no prisons, no limits for subjective creativity! Everything is permitted! Everything is unrestricted!

The Children of Nature receive in joyous ecstasy the chivalrous golden kiss of the Sun and the lascivious, naked, fat belly of the Earth. The Children of Nature springing from the black soil kindle the passions of naked, lustful bodies. They press them all in one spawning, pregnant cup! Thousands of arms and legs are welded into a single suffocating exhausted heap! The skin is inflamed by hot, insatiable, gnawing caresses. Teeth sink with hatred into warm succulent lovers' flesh! Wide, staring eyes follow the pregnant, burning dance of lust! Everything is strange, uninhibited, elemental. Convulsions—flesh—life—death—everything!

Such is the poetry of our love! Powerful, immortal, and terrible are we in our love! The north wind rages in the heads of the Children of Nature. Something frightful has appeared—some vampire of melancholy! Perdition—the world is dying! Catch it! Kill it! No, wait! Frenzied, penetrating cries pierce the air. Wait! Melancholy! Black yawning ulcers of agony cover the pale, terror-stricken face of heaven. The earth trembles with fear beneath the mighty, wrathful blows of its Children! Oh, you cursed, loathsome things! They tear at its fat, tender flesh and bury their withered, starving melancholy in the flowing blood and wounds of its body. The world is dying! Ah! Ah! Ah! cry millions of tocsins. Ah! Ah! Ah! roar the giant cannon of alarm. Destruction! Chaos! Melancholy! The world is dying!

Such is the poetry of our melancholy! We are uninhibited! Not for us the wailing sentimentality of the humanists. Rather, we shall create the triumphant intellectual brotherhood of peoples, forged with the iron logic of contradictions, of Love and Hate. With bared teeth we shall protect our free union, from Africa to the two poles, against any sentimental level of friendship. Everything is ours! Outside us is only death! Raising the black flag of rebellion, we summon all living men who have not been dehumanized, who have not been benumbed by the poisonous breath of Civilization! All to the streets! Forward! Destroy! Kill! Only death admits no return! Extinguish the old! Thunder, lightning, the elements—all are ours! Forward!

Long live the international, intellectual revolution!

An open road for the Anarcho-Furturists, Anarcho-Hyperboreans, and Neo-Nihlists!

Death to world Civilization!

WOMAN AND NATURE:
THE ROARING INSIDE HER (1978)

SUSAN GRIFFIN

He says that woman speaks with nature. That she hears voices from under the earth. That wind blows in her ears and trees whisper to her. That the dead sing through her mouth and the cries of infants are clear to her. But for him this dialogue is over. He says he is not part of this world, that he was set on this world as a stranger. He sets himself apart from woman and nature.

And so it is Goldilocks who goes to the home of the three bears, Little Red Riding Hood who converses with the wolf, Dorothy who befriends a lion, Snow White who talks to the birds, Cinderella with mice as her allies, the Mermaid who is half fish, Thumbelina courted by a mole. *(And when we hear in the Navaho chant of the mountain that a grown man sits and smokes with bears and follows directions given to him by squirrels, we are surprised. We had thought only little girls spoke with animals.)*

We are the bird's eggs. Bird's eggs, flowers, butterflies, rabbits, cows, sheep; we are caterpillars; we are leaves of ivy and sprigs of wallflower. We are women. We rise from the wave. We are gazelle and doe, elephant and whale, lilies and roses and peach, we are air, we are flame, we are oyster and pearl, we are girls. We are woman and nature. And he says he cannot hear us speak.

But we hear.

"WHY CIVILIZATION?"

COMMUNIQUE #23 FROM *DISORDERLY CONDUCT #6*

We are often told that our dreams are unrealistic, our demands impossible, that we are basically out of our fuckin' minds to even propose such a ridiculous concept as the "destruction of civilization." So, we hope that this brief statement may shed some light on why we will settle for nothing less than a completely different reality than what is forced upon us today. We believe that the infinite possibilities of the human experience extend both forwards and backwards. We wish to collapse the discord between these realities. We strive for a "future-primitive" reality, one which all of our ancestors once knew, and one we may come to know: a pre/post-technological, pre/post-industrial, pre/post-colonial, pre/post-capitalist, pre/post-agricultural, and even pre/post-cultural reality—when we were once, and may again be, WILD!

We feel it is necessary to raise some fundamental questions as to where we are now, how we have gotten to this point, where we are headed, and perhaps most importantly, where we have come from. This should not be seen as irrefutable evidence, the Answers, or prescriptions for liberation; but instead, as things to consider while we fight against domination or attempt to create another world.

We believe anarchy to be the ultimate liberatory experience and our natural condition. Before, and outside of, civilization (and its corrupting influences), humans were, and are, for lack of better terms, Anarchistic. For most of our history we lived in small-scale groupings which made decisions face-to-face, without the mediation of government, representation, or even the morality of an abstract thing called culture. We communicated, perceived, and lived in an unmediated, instinctual, and direct way. We knew what to eat, what healed us, and how to survive. We were part of the world around us. There was no artificial separation between the individual, the group, and the rest of life.

In the larger scope of human history, not long ago (some say 10,000 to 12,000 years ago), for reasons we can only speculate about (but never really know), a shift began to occur in a few groupings of humans. These humans began to trust less in the earth as a "giver of life," and began to create a distinction between themselves and the earth. This separation is the foundation of civilization. It is not really a physical thing, although civilization has some very real physical manifestations; but it is more of an orientation, a mindset, a paradigm. It is based on the control and domination of the earth and its inhabitants.

Civilization's main mechanism of control is domestication. It is the controlling, taming, breeding, and modification of life for human benefit (usually for those in power or those striving for power). The domesticating process began to shift humans away from a nomadic way of life, toward a more sedentary and settled existence, which created points of power (taking on a much different dynamic than the more temporal and organic territorial ground), later to be called property. Domestication creates a totalitarian relationship with plants and animals, and eventually, other humans. This mindset sees other life, including other humans, as separate from the domesticator, and is the rationalization for the subjugation of women, children, and for slavery. Domestication is a colonizing force on non-domesticated life, which has brought us to the pathological modern experience of ultimate control of all life, including its genetic structures.

A major step in the civilizing process is the move toward an agrarian society. Agriculture creates a domesticated landscape, a shift from the concept that "the Earth will provide" to "what we will produce from the Earth." The domesticator begins to work against nature and her cycles, and to destroy those who are still living with an understanding her. We can see the beginnings of patriarchy here. We see the beginnings of not only the hoarding of land, but also of its fruits. This notion of ownership of land and surplus creates never-before-experienced power dynamics, including institutionalized hierarchies and organized warfare. We have moved down an unsustainable and disastrous road.

Over the next thousands of years this disease progresses, with its colonizing and imperialist mentality eventually consuming most of the planet with, of course, the help of the religious-propagandists who try to assure the "masses" and the "savages" that this is good and right. For the benefit of the colonizer, peoples are pitted against other peoples. When the colonizer's words do not suffice, the sword is never far away with its genocidal collision. As the class distinctions become more solidified, there becomes only those who have, and those who do not. The takers and the givers. The rulers and the ruled. The walls get raised. This is how we are told it has always been; but most people somehow know this isn't right, and there have always been those who have fought against it.

The war on women, the war on the poor, the war on indigenous and land-based people, and the war on the wild are all interconnected. In the eyes of civilization, they are all seen as commodities—things to be claimed, extracted, and manipulated for power and control. They are all seen as resources; and when they are of use no longer to the power structure, they are discarded into the landfills of society. The ideology of patriarchy is one of control over self-determination and sustainability, of reason over instinct and anarchy, and of order over freedom and wildness. Patriarchy is an imposition of death, rather than a celebration of life. These are the motivations of patriarchy and civilization; and for thousands of years they have shaped the human experience on every level, from the institutional to the personal, while they have devoured life.

The civilizing process became more refined and efficient as time went on. Capitalism became its mode of operation, and the gauge of the extent of domination and the measure of what still is needed to be conquered. The entire planet was mapped and lands were enclosed. The nation-state eventually became the proposed societal grouping, and it was to set forth the values and goals of vast numbers of peoples, of course, for the benefit of those in control. Propaganda by the state, and the by now less powerful church-started to replace some (but certainly not most) of the brute force with on-the-surface benevolence and concepts like citizenry and democracy. As the dawn of modernity approached, things were really getting sick.

Throughout the development of civilization, technology always played an ever-expanding role. In fact, civilization's progress has always been directly connected to, and determined by, the development of ever more complex, efficient, and innovative technologies. It is hard to tell whether civilization pushes technology, or vice-versa. Technology, like civilization, can be seen more as a process or complex system than as a physical form. It inherently involves division of labor, resource extraction, and exploitation by power (those with the technology). The interface with, and result of, technology is always an alienated, mediated and heavily-loaded reality. No, technology is not neutral. The values and goals of those who produce and control technology are always embedded within it. Different from simple tools, technology is connected to a larger momentum. This technological system always advances, and always needs to be inventing new ways to support, fuel, maintain, and sell itself. A key part of the modern-techno-capitalist structure is industrialism; the mechanized system of production built on centralized power, and the exploitation of people and nature. Industrialism cannot exist without evictions, forced labor, cultural destruction, assimilation, ecological devastation, and global trade are accepted and seen as necessary. Industrialism's standardization of life objectifies and commodifies it, viewing all life as a potential resource. Technology and industrialism have opened the door to the ultimate domestication of life—the final stage of civilization— the age of neo-life.

So now we are in the post-modern, neo-liberal, bio-tech, cyber-reality, with an apocalyptic future and new world order. Can it really get much worse? Or has it always been this bad? We are almost completely domesticated, except for the few brief moments (riots, creeping through the dark to destroy machinery or civilization's infrastructure, connecting with other species, swimming naked in a mountain stream, eating wild foods, lovemaking... add your own favorites) when we catch a glimpse of what it would be like to go feral. Their "global village" is more like a global amusement park or global zoo, and it's not a question of boycotting it 'cause we're all in it, and it's in all of us. And we can't just break out of our own cages (although we're helpless unless we start there), but we gotta bust down the whole fuckin' place, feast on the zookeepers and those who run and benefit from it, reconnect with our own instincts, and become wild again! We cannot reform civilization, green it up, or make it more fair. It is rotten to the core. We don't need more ideology, morality, fundamentalism, or better organization to save us. We must save ourselves. We have to live according to our own desires. We have to connect with ourselves, those we care about, and the rest of life. We have to break out of, and break down, this reality.

We need Action.

To put it simply, civilization is a war on life.

We are fighting for our lives, and we declare war on civilization!

Under the pavement, the beach.

—Paris, 1968

Sources

Section I. Outside Civilization

Hoxie Neal Fairchild: from *The Noble Savage* by Hoxie Neal Fairchild. Copyright © 1928 by Columbia University Press. Reprinted with permission of the publisher.

Henry David Thoreau: "Excursions" in Charles R. Murphy, ed., *Little Essays from the Works of Henry David Thoreau*

Marshall Sahlins: "The Original Affluent Society" in Richard B. Lee and Irven De Vore, eds., *Man the Hunter*

Lynn Clive: "Birds Combat Civilization" in *Fifth Estate*, Summer 1985

John Landau: "Wildflowers: A Bouquet of Theses" in Primal Revival Growth Center, Los Angeles, 1998

Theodor Adorno: "Minima Moralia: Reflections from Damaged Life" by Theodor Adorno. Copyright © 1974 by Verso Press. Reprinted by permission of the publisher.

Marvin Harris, *Our Kind: Who We Are, Where We Came From, Where We are Going*, Harper and Row, 1989

Ramona Wilson, "Spokane Museum," from *Dancing on the Rim of the World: An Anthology of Contemporary Northwest Native American Writing*, edited by Andrea Lerner, Sun Tracks and The University of Arizona Press, Tucson, 1990

Section II. The Coming of Civilization

Frederick Turner: "Beyond Geography: The Western Spirit Against the Wilderness" from Frederick W. Turner, III: from *Beyond Geography*. Copyright © 1980 by Frederick Turner. Used by permission of Viking Penguin, a division of Penguin Putnam Inc.

James Axtell: from *The Invasion Within: The Contest of Cultures in Colonial North America* by James Axtell. Copyright © 1985 by James Axtell. Used by permission of Oxford University Press, Inc.

Mark Nathan Cohen: "Health and the Rise of Civilization." Copyright © 1989 by Mark Nathan Cohen. Used by permission of Yale University Press.

Chellis Glendinning: from *My Name is Chellis and I'm in Recovery from Western Civilization* by Chellis Glendinning. Copyright © 1994. Reprinted by arrangement with Shambhala Publications, Inc., 300 Massachusetts Avenue, Boston, MA 02115.

Pierre Clastres: *Society Against the State: Essays in Political Anthropology*, Zone Books, 1987

Madhusree Mukerjee: from *The Land of the Naked People: Encounters with Stone Age Islanders*, Houghton Mifflin, 2003

Robert Wolff: "Reading and Writing" From *Original Wisdom: Stories of an Ancient Way of Knowing*. Rochester, Vermont: Inner Traditions, 2001. Copyright © Robert Wolff.

SECTION III. THE NATURE OF CIVILIZATION

Charles Fourier: "Theory of Four Movements and General Destinies" in *Oeuvres Complètes*

Sigmund Freud: from *Civilization and Its Discontents* by Sigmund Freud, translated by James Strachey. Translation copyright © 1961 by James Strachey, renewed 1989 by Alix Strachey. Reprinted by permission of W.W. Norton & Company, Inc.

John Landau: "Civilization and the Primitive" (Unpublished, 1995)

Richard Heinberg: "Was Civilization a Mistake?" in *Green Anarchist*, Autumn 1997

Zygmunt Bauman: reprinted from *Modernity and the Holocaust*. Copyright © 1989 by Zygmunt Bauman. Used by permission of the American publisher, Cornell University Press, and by Blackwell Publishers, Oxford, U.K.

T. Fulano: "Civilization Is Like a Jetliner" in *Fifth Estate*, Winter 1983

Unabomber (AKA "FC"): "Industrial Society and Its Future" in *Washington Post*, September 21, 1995

Tamarack Song: from *The Old Way and Civilization*, Station Hill Press, 1994

SECTION IV. THE PATHOLOGY OF CIVILIZATION

Joseph A. Tainter: from *The Collapse of Complex Societies* by Joseph Tainter. Copyright © 1988 by Joseph A. Tainter. Reprinted by permission of Cambridge University Press.

Andrew Bard Schmookler: "The Parable of the Tribes: The Problem of Power in Social Evolution." Reprinted from *The Parable of the Tribes* by Andrew Bard Schmookler by permission of the State University of New York Press. Copyright © 1995 by Andrew Bard Schmookler.

Peter Sloterdijk: from *Critique of Cynical Reason* by Peter Sloterdijk, translated by Michael Eldred. Copyright © 1987 by Peter Sloterdijk. Reprinted by permission of University of Minnesota Press.

Frederic Jameson: from *The Seeds of Time* by Frederic Jameson. Copyright © 1994 by Columbia University Press. Reprinted with permission of the publisher.

labor of ludd: "The Medium Is the Medium" from Dan Todd poster, Tucson, 1998

Des Réfractaires: "How Nice to Be Civilized!" in *Anarchy*, Summer 1993

David Watson: "Civilization in Bulk" in *Fifth Estate*, Summer 1991

Chrystos: "They're Always Telling Me I'm Too Angry" from *Fugitive Colors*, 1995, Cleveland State University Poetry Center

Oswald Spengler, *Man and Technics: A Contribution to a Philosophy of Life*, translated from the German by Charles Francis Atkinson, Knopf, 1931

John Mohawk: "In Search of Noble Ancestors" from *Civilization in Crisis: Anthropological Perspectives*, Edited by Christine Ward Galley, 1992

SECTION V. THE RESISTANCE TO CIVILIZATION

Rudolf Bahro: from *Avoiding Social and Ecological Disaster: The Politics of World Transformation*. Copyright © 1994 by Gateway Books, Bath, U.K. Used by permission of the publisher.

William Morris: from *Morris: News from Nowhere* by William Morris, edited by Krishan Kumar. Copyright © 1995 by Cambridge University Press. Reprinted by permission of Cambridge University Press.

Feral Faun: "Feral Revolution" in *Demolition Derby* #1, 1988

Anonymous: "Don't Eat Your Revolution! Make It!" in *News & Views From (the former) Sovietsky-Soyuz*, February 1995

Glenn Parton: "The Machine in Our Heads" in *Green Anarchist*, Summer 1997

Alon K. Raab: "Revolt of the Bats" in *The Bear Essential*, Summer 1995

Kirkpatrick Sale: "Rebels Against the Future: Lessons from the Luddites" from Kirkpatrick Sale, *Rebels Against the Future*. Copyright © 1995 Kirkpatrick Sale. Reprinted by permission of Addison-Wesley Longman Inc.

Derrick Jensen: "Actions Speak Louder Than Words" in *Earth First! Journal*, May-June 1998

Anti-Authoritarians Anonymous: "We Have To Dismantle All This" in Anti-Authoritarians Anonymous flyer, Eugene, OR 1995

John (Fire) Lame Deer and Richard Erdoes: "Talking to the Owls and Butterflies" from *Lame Deer: Seeker of Visions* (Washington Square Press, 1976)

Group of Anarcho-Futurists: Anarcho-Futurist Manifesto K. Svetu's Shturmovol, opustosshalushchii manifest anarkho-furturistov (Kharkov, 3/14/1919). From Paul Avrich's *The Anarchists in the Russian Revolution* (Thames & Hudson, 1973)

Susan Griffin: from *Woman and Nature: The Roaring Inside Her*, Harper Colophon Books, 1978

Also From Feral House

WAR IS A RACKET
The Antiwar Classic by America's Most Decorated General
By General Smedley D. Butler • Introduction by Adam Parfrey

General Butler's notorious 1933 speech, "War is a Racket," excoriates "the small inside group" that "knows what the racket is all about." The Feral House edition provides more rare anti-imperialist screeds and an exposé of a Congressional inquiry into Butler's whistleblowing of a coup d'état attempt by big business against Franklin Delano Roosevelt.

5 x 8 • 84 pages • paperback original • ISBN: 0-922915-86-5 • $9.95

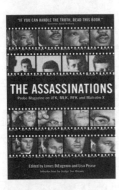

THE ASSASSINATIONS
Probe Magazine on JFK, MLK, RFK, and Malcolm X
Edited by James DiEugenio and Lisa Pease

"Throughout the '90s, Lisa Pease and Jim DiEugenio covered the issues and reported on new developments in the JFK, MLK and RFK assassination cases in remarkable depth. Their book is a must-read to understand how these leaders were systematically eliminated. It is a unique volume in the literature on this subject."
—William Turner, former FBI agent and co-author of *Deadly Secrets*

6 x 9 • 684 pages • paperback • ISBN: 0-922915-82-2 • $24.00

RUNNING ON EMPTINESS
The Pathology of Civilization
By John Zerzan

"John Zerzan is the most important philosopher of our time. All the rest of us are building on his foundation. His unrelenting questions and careful analyses point us toward the necessary goal: the unmaking of civilization."
—Derrick Jensen, author of *A Language Older Than Words* and *Listening to the Land*

"Today's pundits are quick to assume the contrarian mantle, but John Zerzan does the hard work to earn it. He runs deeply against the tide of familiar arguments from the left and right. Pay attention to his wake—you'll find your definition of 'liberty' suddenly expanding."
—James MacKinnon, senior editor, *Adbusters*

5 1/2 x 8 1/2 • Paperback original • 214 pages • ISBN: 0-922915-75-X • $12.00

TO ORDER FROM FERAL HOUSE:
Individuals: Send check or money order to Feral House, P.O. Box 39910, Los Angeles CA 90039, USA. For credit card orders: call (800) 967-7885 or fax your info to (323) 666-3330. CA residents please add 8.25% sales tax. U.S. shipping: add $4.50 for first item, $2 each additional item. Shipping to Canada and Mexico: add $9 for first item, $6 each additional item. Other countries: add $11 for first item, $9 each additional item. Non-U.S. originated orders must include international money order or check for U.S. funds drawn on a U.S. bank. We are sorry, but we cannot process non-U.S. credit cards.
www.feralhouse.com